Henry Leffmann

Examination of Water for Sanitary and Technic Purposes

Henry Leffmann

Examination of Water for Sanitary and Technic Purposes

ISBN/EAN: 9783743686984

Printed in Europe, USA, Canada, Australia, Japan

Cover: Foto ©berggeist007 / pixelio.de

More available books at **www.hansebooks.com**

EXAMINATION OF WATER

FOR

SANITARY AND TECHNIC PURPOSES.

BY

HENRY LEFFMANN, A.M., M.D., Ph.D.,

PROFESSOR OF CHEMISTRY IN THE WOMAN'S MEDICAL COLLEGE OF PENNSYLVANIA, IN THE PENNSYLVANIA COLLEGE OF DENTAL SURGERY, AND IN THE WAGNER FREE INSTITUTE OF SCIENCE; PATHOLOGICAL CHEMIST TO THE JEFFERSON MEDICAL COLLEGE HOSPITAL; CHEMIST TO DAIRY AND FOOD COMMISSIONER OF PA.

THIRD EDITION, REVISED AND ENLARGED, WITH ILLUSTRATIONS.

PHILADELPHIA:
P. BLAKISTON, SON & CO.,
1012 WALNUT STREET.
1895.

Copyright, 1895, by P. BLAKISTON, SON & CO.

PRESS OF WM. F. FELL & CO.,
1220-24 SANSOM STREET,
PHILADELPHIA.

I DEDICATE THIS BOOK TO THE

MEMORY OF

MY MOTHER,

TO WHOSE WISE PRECEPT AND EXAMPLE

IN MY BABYHOOD

I OWE WHATEVER MERIT MY MANHOOD YEARS

MAY SHOW.

PREFACE TO THIRD EDITION.

In the present edition the general character of the work as heretofore exhibited has been preserved, but numerous modifications of the details of processes of sanitary analysis as developed by various American workers have been described. Much valuable matter has of late years been contributed by the various State stations for sanitary control, and while these methods have not solved the difficult problem presented to the sanitary chemist, namely, the absolute judgment as to whether a given water-sample is or is not wholesome, they have aided in securing uniformity of results, and the accumulation of data upon which judgment may be formed. Much of the text in the section on biologic examinations has been re-written, and I have taken occasion to express more at length the sources of uncertainty in this field; sources which were recognized and pointed out in both the earlier editions, but which have become much more appreciable to the profession at large within the last few years. Whatever may be the shortcomings of the standard analytic examinations, it seems probable that for some years to come, at least, the results will remain the most satisfactory basis for judgment as to the potability of a water. There can be no question that many of the published results as to bacteriologic examinations of water-samples are without value, partly from the inherent difficulties of the methods, even in the most expert and honest hands; partly, it is to be regretted, from absolute misstatement. I have had personal knowledge of one instance of the latter phase.

My former colleague, Dr. William Beam, now permanently residing in a distant part of the world, has asked to be relieved from his share in the work. The copyright privileges have been formally transferred to me, and I have conducted entirely the work of the present revision.
H. L.

715 Walnut St.
Philadelphia, August 1895.

PREFACE TO SECOND EDITION.

In the period that has elapsed since the publication of the first edition of this work, many processes for water analysis have been proposed, and we have included in the present revision such of these as seem to us of substantial value. In particular we may mention the methods recommended by the Chemical Section of the American Association for the Advancement of Science, and the application of the Kjeldahl process to the determination of the organic nitrogen. The adoption of the former methods will serve to secure uniformity in analytical data, while it is to be hoped that chemists generally will investigate and report on the latter, in order that a basis for the interpretation of results may be obtained.

No material change has been made in the description of the general Quantitative Analysis, in which we have followed to a large extent the methods indicated by Fresenius, selecting those best adapted to technical purposes.

We have extended considerably the section on the Biological Examinations, although we have seen no reason to change the opinions expressed in the former edition as to the value of these results. It would be impossible to overestimate the importance of bacteriology, in etiology, pathology, and general biology, but until pathogenic microbes are more clearly indicated and described, the methods will be of little use in dealing with the problem of the determination of the sanitary and technical value of water supplies.

In the chapter on the Purification of Water we have described in some detail a few of the important systems, especially the Anderson iron process, the efficiency of which we have had ample opportunity to observe by experiments on a comparatively large scale, extending over several months.

The favorable reception accorded to the first edition, both by journals of acknowledged authority, and by chemists of extended experience in this department, has indicated that the work is not without usefulness in the field to which it is devoted.

H. L.
W. B.

CONTENTS.

                                                                                        PAGES

HISTORY AND CLASSIFICATION OF NATURAL WATERS.
    Rain Water—Surface Water—Subsoil Water—Deep Water, ... 13–20

ANALYTIC OPERATIONS.
    Sanitary Examinations:—
        Collection and Preliminary Examination—Total Solids—Chlorin—Nitrogen in Ammonium Compounds and Organic Matter—Nitrogen as Nitrates —Nitrogen as Nitrites—Oxygen-consuming Power — Phosphates — Dissolved Oxygen — Poisonous Metals—Biologic Examinations, ... 21–74

    Technic Examinations:—
        General Quantitative Analysis—Spectroscopic Analysis—Specific Gravity, ... 74–88

INTERPRETATION OF RESULTS.
    Statement of Analysis—Sanitary Applications—Action of Water on Lead, ... 89–109

TECHNIC APPLICATIONS.
    Boiler Waters—Sewage Effluents—Purification of Water—Identification of Source of Water, . . . 109–135

ADDENDA, ... 136–140

ANALYTIC DATA.
    Factors for Calculation—Conversion Table—Oxygen Dissolved in Water—Rain and Subsoil Waters— Schuylkill Water—Artesian Waters—City Supplies —Culture Phenomena, ... 141–149

INDEX, ... 151–154

CORRECTIONS.

On page 137, lines 6 and 5 from bottom should read "nutrient agar-agar of definite reaction (see below for methods of determining reaction), to which 2 to 3 per"

On same page, line 2 from bottom, for *typhus* read *typhi*.

EXAMINATION OF WATER.

HISTORY AND CLASSIFICATION OF NATURAL WATERS.

Pure water is an artificial product. Natural waters always contain foreign matters in solution and suspension, varying from mere traces to very large proportions. The properties, effects, and uses of water are considerably modified by these ingredients, and the object of analysis is to ascertain their character and amount. Since these are largely dependent on the history of the water, a classification based on this will be convenient. We may distinguish four classes of natural waters :—

Rain Water.—Water precipitated from the atmosphere under any conditions, and therefore including dew, frost, snow, and hail.

Surface Water.—All collections of water in free contact with the atmosphere, as in streams, seas, lakes, or ponds.

Subsoil or *Ground Water.*—Water not in free contact with the atmosphere, percolating or flowing through soil or rock at moderate distance below the surface, and derived in large part from the rain or surface water of the district.

Deep or *Artesian Water.*—Water accumulated at considerable depth below the surface, from which the subsoi water of the district has been excluded by difficultly permeable strata.

Rain Water, when gathered in the open country and in the later period of a long rain or snow, is the purest form of natural water. When collected directly, it contains but little solid matter, this consisting principally of ammonium compounds and particles of organic matter, living and dead, gathered from the atmosphere. In districts near the sea an appreciable amount of chlorids will be present. It is obvious that a prolonged rain will wash out the air, but since storms are usually attended by wind, fresh portions of air are continually flowing in, and thus the water never becomes perfectly pure. Rain water collected in inhabited districts is usually quite impure.

Surface Water.—Rain water in part flows off on the surface, and gains in the proportion of suspended and dissolved matters, the former being found in large amount when the rainfall is profuse. The wearing action of water is dependent on the amount and character of these suspended materials. From the higher levels of a watershed, the streams, more or less in the form of torrents, gather into larger currents, and reaching lower levels become slower in movement, and deposit much of the suspended matter. By admixture of the waters from widely separated districts the character and amount of the dissolved matters are much modified. An action of this kind is seen in the watershed of the Schuylkill River. This stream rises in the anthracite-coal region of Pennsylvania, and receiving much refuse mine-water becomes impregnated with iron salts and free mineral acid, being then quite unsuitable for drinking or manufacturing purposes. In its course of about one hundred miles it passes over an extensive limestone district, and receives several large streams highly charged with calcium carbonate. The result is a neutralization of the acid and a precipitation of the iron and much of the calcium. The river becomes purer, and at its junction with the Delaware River at Philadelphia it contains neither free sulphuric nor hydrochloric acid, only traces of iron, and but a small amount of calcium sulphate. In this manner there is produced a soft water, superior to that of the river near its source, or to the hard waters of the middle Schuylkill region.

It is obviously impossible to establish close standards of composition for surface waters. In the case of rain water, falling on the surface of undisturbed, unpopulated territory, the amount of solids dissolved will be small, and will consist principally of carbonates and sulphates. The water of lakes and rivers is, however, in part derived from springs, which may proceed from great depths, and thus introduce substances not easily soluble in surface water, nor derivable from the soil of the district.

The exposure to light and air which surface water undergoes, results in the absorption of oxygen and loss of carbonic acid, together with the oxidation of the organic matter. The diminution of the rapidity of the current permits the deposition of the suspended matters, and this occurs especially as the river approaches the sea, not only from the retarding influence of the tidal wave, but from the precipitating action of the salt water. The investigations of Carl Barus, published in Bulletin No. 36, U. S. Geological Survey, have shown the decided influence of sodium chlorid in accelerating the subsidence of fine particles.

Subsoil Water.—Water that penetrates the soil,

passes to various depths, according to the porosity and arrangement of the strata. As a rule, it descends until it reaches but slightly pervious formations, upon the level of which it accumulates. In the upper layer of soil it dissolves mineral and organic ingredients, and becomes impregnated with microörganisms, through the agency of which the organic matter undergoes important transformations. The water constantly accumulating, gradually flows along the incline of the impervious stratum, or through its fissures, and may either pass downward or emerge in the form of a spring.

The proportion of water which may be held by any rock or soil is often much larger than would be at first supposed. T. Sterry Hunt states that a square mile of sandstone 100 feet thick will contain water sufficient to sustain a flow of a cubic foot a minute for more than thirteen years.

Much difference is observed in the composition of subsoil waters, but as a general rule they contain but a limited amount of mineral substances, and a very small proportion of organic matter. In populated districts, however, a marked change is produced through admixture with water containing animal and vegetable products in various stages of decomposition. It is especially the organic matter containing nitrogen that is of importance. To this class belong all those compounds forming tissues that are intimately associated with vital action; also many characteristic excretory products. These bodies are mostly unstable, and as soon as their vitality ceases begin to decompose, partly by oxidation, partly by splitting up into simpler forms; these changes being in most cases brought about by microörganisms. Among the products noticed in the early stages of such decay, are substances which possess close analogies to the organic bases or alkaloids, but more susceptible of decomposition. They are generally present in minute amount, but are not infrequently very active in their physiological effect. From the most recent researches it seems probable that the pathogenic power of many microörganisms rests not upon any mechanical or other action of the germs themselves, but upon the alkaloidal principles which they produce and excrete. As a group these bodies are known as the "ptomains." Nitrogen is an invariable ingredient. The ultimate results of the processes of decomposition depend largely on circumstances. When organic matters containing nitrogen are subjected to the action of oxidizing agents, such as alkaline potassium permanganate or chromic acid, some of the nitrogen is converted into ammonium compounds. A similar result occurs in all waters, but a considerable portion of the organic matter may also suffer further oxidation, and in association with the mineral substances present form nitrites and nitrates, especially the latter. This conversion is called "nitrification." The conditions under which it occurs have been carefully studied by various observers, but much research is still needed to place the subject on a secure basis.

Nitrification takes place under the influence of microbes, the habitat of which does not extend more than a few yards below the surface of the soil. Percy and Grace Frankland have isolated and described a bacillus with active nitrifying powers. It is a very short, almost spherical form, which grows in ammoniacal culture-fluids and in meat-broth, but does not grow in the usual gelatin-peptone mixture, except when previously cultivated in meat-broth. The nitrifying action is probably exerted only upon the ammonium which is formed from the organic matter. The presence of some substance capable of neutralizing acids is necessary to continuous action. Calcium and magnesium carbonates fulfill this function. Nitrates are the final result of this action ; nitrites are present at any given time only in small quantity. Denitrification, that is, the reduction of nitrates and nitrites to ammonium compounds, takes place also under the influence of microbes, and is especially apt to occur when considerable quantities of decomposing organic matter are introduced. Percy and Grace Frankland have described several species of bacilli which have active denitrifying powers. Among these are *Bacillus liquidus*, *B. vermicularis*, and *B. ramosus*. A partial reduction sometimes occurs, and a notable proportion of nitrites is found, but in the presence of actively decomposing organic matter, such as that in sewage, a complete reduction, even to the liberation of nitrogen, may occur. Jordan (*Rep. S. B. of H., of Mass.*, 1890), has described several species, found in the sewage of the city of Lawrence, which reduce nitrates rapidly to nitrites.

Deep Water.—Water which penetrates the fissures of the fundamental rock formations may pass to great depths, and by following the lines of the lowest and least permeable strata may be transported to points far removed from those at which it was originally collected. The chemical changes thus induced include most of those which take place at higher points, but the increase of pressure and temperature confers increased solvent power. Carbonic acid will accumulate under conditions favorable to the solution of calcium, magnesium, and iron carbonates, and iron and manganese oxids may be converted into carbonates and then dissolved. Sulphates are reduced to sulphids, and these subsequently, by the action of carbonic acid, yield hydrogen sulphid. Organic matter, living and dead, plays an important part, determining the reduction of ferric compounds to ferrous, and of the sulphates to sulphids, and is itself converted ultimately into ammonium compounds, notable quantities of which are often found in deep waters. Further, it is found that nitrates and nitrites are present only in small amount, except from certain strata rich in organic matter. In some cases the water acquires very high temperature, and dissociation of rocks occurs with solution of considerable amounts of silicic acid, which is ordinarily but sparingly soluble in water.

Masses of water thus accumulated under heat and pressure may find their way to the surface either through natural fissures, or be reached by borings. The mineral springs, highly charged with solid matters, and the artesian waters are obtained in this way.

While no absolute unchangeable line can be drawn between deep and subsoil waters, yet it will in most cases be found that the deep water of a given district, whether obtained through natural or artificial channels, will be decidedly different in composition from the subsoil or surface water of the same, and that the rocks passed through in such cases will be characterized by one or more strata, difficultly permeable to water, and therefore preventing direct communication. The characteristic differences between surface, subsoil, and deep waters are clearly indicated in the table of analyses given in the appendix.

The fact that mere depth is not the essential difference between the two classes of waters is shown by comparison between the composition of the water from the well at Barren Hill, on the northern border of Philadelphia county, and the deep well at Locust Point, Baltimore. The former is a dug well, 130 feet deep; the latter is an artesian boring of 128 feet, which in its descent passes through four feet of solid rock. The deeper well is evidently supplied by subsoil water. The artesian well, though located 100 yards from a brackish sewage-laden estuary, evidently derives no water from it.

ANALYTIC OPERATIONS.

SANITARY EXAMINATIONS.

COLLECTION AND PRELIMINARY EXAMINATION OF SAMPLES.

Great care must be taken in collecting water-samples, in order to secure a fair representation of the supply and to avoid introduction of foreign matters. The five-pint green glass stoppered bottles used for holding acids are suitable for containing the samples. The contents of one such bottle will suffice for most sanitary or technical examinations. Fig. 1 shows a boxed bottle known as "Banker's Glass Can," which I have found very convenient for transportation. It is provided with a hinged lid which can be fastened, if deemed necessary, by a padlock. The green-glass stoppered bottles may be fitted in such an arrangement. Stone jugs, casks, or metal vessels must not be employed. The bottles used must be

FIG. 1.

thoroughly rinsed several times with the water to be examined, filled and the stopper tied down, or fastened by stretching a rubber finger-cot over the stopper and lip. If corks are used, they should be new and thoroughly rinsed. Wax, putty, plaster, or similar material should not be used.

In taking samples from lakes, slow streams, or reservoirs, it is necessary to submerge the bottle so as to avoid collecting any water that has been in immediate contact with the air.

In the examination of public water supplies, the sample should be drawn from a hydrant in direct connection with the main, and not from a cistern, storage tank, or dead end of a pipe. In the case of pump-wells, a few gallons of water should be pumped out before taking the sample, in order to remove that which has been standing in the pipe.

In all cases care should be taken to fill the vessel with as little agitation with air as possible.

It is important that with each sample a record be made of those surroundings and conditions which might influence the character of the water, particularly in reference to sources of pollution, such as proximity to cesspools, sewers, or manufacturing establishments. The character and condition of the different strata of the locality should be noted if possible.

Determinations of nitrogen existing as ammonium compounds, and as organic matter, and of oxygen-consuming power, should be made upon the sample in the original condition, whether turbid or clear, but all other estimations should be made upon the clear liquid. Turbid waters may be clarified by standing or by filtration; for the latter purpose Schleicher & Schüll's extra-heavy No. 598 paper is the best. In many cases the suspended matter cannot be entirely removed by filtration, and subsidence must be resorted to. The use of a small quantity of alum, or aluminum hydroxid, as described in the section on the purification of water, will sometimes be applicable as a means of clarifying samples. For the quantitative determination, the sediment from a known volume of the water is collected on a tared filter, dried at 112° F., and weighed.

The water from newly-dug wells is generally turbid and the determinations are best made after filtration, but the results will be unsatisfactory, showing a higher proportion of organic matter than will be found when the supply becomes clear.

The following methods of determining color and odor have been adopted by the Society of Public Analysts of Great Britain:—

Color.—A colorless glass tube, two feet long and two inches in diameter, is closed at each end with a disc of colorless glass. An opening for filling and emptying the tube should be made at one end, either by cutting a small segment off the glass disc, or cutting out a small segmental section of the tube itself before the disc is cemented on. A good cement for such purposes is the following:—

Caoutchouc,	2 parts.
Mastic,	6 "
Chloroform,	100 "

The ingredients are mixed and allowed to stand for a few days. The cement should be used as soon as solution is effected, as it becomes viscid on standing.

The tube must be about half filled with the water to be examined, brought into a horizontal position, level with the eye, and directed toward a brightly illuminated white surface. The comparison of tint has to be made between the lower half of the tube containing the water under examination, and the upper half containing air only.

By determining the amount of ammonium chlorid required to produce with a standard quantity of Nessler solution a color equal in depth to that of the water-sample, a means of recording the color with some precision may often be obtained. The manipulation is conducted as described in connection with the determination of ammonium compounds, using equal volumes of the sample and of distilled water in separate cylinders. The Nessler solution is, of course, added only to the distilled water. The volume of diluted standard ammonium chlorid solution which must be added to the distilled water to enable it to produce with the Nessler reagent a color equal to that of the sample is taken as an index of the color of the latter. Thus "color = 0.3" means that the color of the sample equals that produced by Nessler solution in distilled water to which 0.3. c. c. of the diluted standard ammonium chlorid solution has been added.

Odor.—Put about 150 c. c. of the water into a clean, wide-mouth 250 c. c. stoppered bottle, which has been previously rinsed with the same water; insert the stopper and warm the water in a water-bath to 100° F. Remove the bottle from the water-bath and shake it rapidly for a few seconds; remove the stopper and immediately note if the water has any smell. Insert the stopper and repeat the test.

In a polluted water the odor will sometimes give a clue to the origin of the pollution.

Reaction.—The determination of reaction is usually

made by the addition of a neutral solution of litmus to the water. If an acid reaction is obtained the water should be boiled in order to determine if it is due to carbonic acid. Some of the more delicate indicators, such as phenolphthaleïn and lacmoid, may be used with advantage for these tests. The latter possesses the advantage that it is unaffected by carbonic acid, but detects even traces of free mineral acid. It is neutral, also, to many normal metallic salts, such as ferrous sulphate, which are acid to litmus. Ferric salts, however, are acid to lacmoid. Its color changes are the same as those of litmus, *i. e.*, red with acids and blue with alkalies.

Phenolphthaleïn is best applied to the detection of weak acids, such as carbonic acid and the organic acids. In acid and neutral solutions it is colorless—in alkaline, deep red. Nearly all waters contain carbonic acid, and will therefore bleach a solution of phenolphthaleïn which has been reddened by a small amount of alkali.

TOTAL SOLIDS.

A platinum basin holding 100 c. c. will be found convenient for this determination. This should weigh about 45 grams. It should be kept clean and smooth by frequent burnishing with sand, a little of which should be placed in the palm of the hand, moistened, and the dish gently rubbed against it. Very fine sea-sand with round, smooth grains, is the only kind suitable for this purpose. Coarse river-sand, tripoli, or other rough scouring-powders, must not be employed. If proper care is taken, the lustre of the metal will remain unimpaired indefinitely, and the loss in weight will be trifling. The inner surface can generally be cleaned by treatment with hydrochloric acid, rinsing and, if necessary, burnishing. Neglect of these precautions will soon lead to serious damage to the dish. A small, smooth slab of iron or marble is convenient to set it on while cooling. When being heated over the naked flame the dish should rest on a triangle of iron wire, covered with pipe-stems. Dishes of pure nickel are not satisfactory substitutes for those of platinum.

Platinum-pointed forceps should be used in handling the dish. The platinum terminals may be kept bright and clean by the use of sand.

Fig. 2.

The low-temperature burner, used as shown in Fig. 2, will be found a very convenient substitute for the water-bath and hot air oven. The inlet pipe is very short and soon becomes so hot as to injure the rubber tube. To avoid this it may be lengthened by means of a piece of $\frac{1}{8}$-inch gas-pipe, or the junction may be wrapped with a rag, the ends of which dip into water. By capillary attraction the rag is kept moist and cool.

The determination of total solids is made by evaporating 100 or 200 c. c. of the water in the platinum basin, which has been previously heated almost to redness, allowed to cool for ten minutes, and weighed. The operation is conducted at a moderate heat. When the residue appears dry, the heat may be increased slightly for some minutes. The above method will answer in most cases. In waters of exceptional purity it may be advisable to use larger quantities, such as 250 c. c. When the residue contains deliquescent bodies, the determination will not be accurate, and when appreciable amounts of magnesium and chlorin are present, a decomposition will occur toward the close of the evaporation by which magnesium oxid will be formed and hydrogen chlorid escape.

The irregular decomposition occurring during the evaporation may be largely prevented by adding .005 gm. of sodium carbonate to each 100 c. c. of the sample taken. This converts magnesium and calcium salts into carbonates. The sodium carbonate is conveniently kept in the form of solution of such strength that 1 c. c. contains .001 gm. The weight of the carbonate is of course to be deducted from the weight of the residue. Drown and Hazen have carefully investigated this method and have found it available for a more satisfactory determination of the loss on ignition. For this process they place the platinum basin containing the residue within another similar basin of such size that an air-space of about one-half inch is left all around the inner dish, which is supported upon a spiral of platinum that rests on the bottom of the outer dish. Over the inner dish is suspended a disc of platinum foil to reflect the heat. The outer dish is heated to bright redness.

After the weight of the residue is obtained, the dish should be cautiously heated to low redness, and the effect noted. Nitrates and nitrites, calcium and magnesium carbonates are decomposed; ammonium salts are driven off; potassium and sodium chlorids are also driven off if the temperature is high. Organic matter is at first charred, and by continued heating burned off. When the quantity of nitrates is considerable, slight deflagration may be observed, or the production of red fumes of nitrogen dioxid. The organic matter, in decomposing, not infrequently develops odors which indicate its character or source. These are more satisfactorily observed when a rather large quantity, say 250 c. c., is evaporated at a low heat, preferably on a water-bath.

In water of high organic purity, the residue on heating will give no appreciable blackening nor odor, while in forest streams charged with vegetable matter derived from falling leaves, very decided blackening without unpleasant odor will be noticed. The loss of weight after heating cannot be taken as a measure of the organic matter, except when present in relatively large amount.

CHLORIN.

Solutions Required :—

Standard Silver Nitrate.—Dissolve about five grams of pure recrystallized silver nitrate in distilled water, and make the solution up to 1000 c. c. The amount of chlorin to which this is equivalent may be determined as follows: Several grams of pure sodium chlorid are finely powdered and heated over a Bunsen burner for five minutes, not quite to redness. When cold, 0.824 gram is dissolved in water and the solution made up to 500 c. c. 25 c. c. of this should be treated as below, and the amount of silver solution required noted. Each c. c. of the sodium chlorid solution is equivalent to .001 gram chlorin.

Potassium Chromate.—Five grams of potassium chromate are dissolved in 100 c. c. of distilled water. A solution of silver nitrate is added until a permanent red precipitate is produced, which is separated by filtration.

Analytic Process :—

If a preliminary test shows the chlorin to be present in considerable amount, the determination may be made on

100 c. c. of the water without concentration. If, however, there is but little present, 250 c. c. should be evaporated to about one-fifth, best with the addition of a little sodium carbonate, and the determination made on the concentrated liquid after cooling.

The water is placed in a porcelain dish or in a beaker standing on a white surface, a few drops of potassium chromate solution added, and standard silver nitrate solution run in from a burette until a faint red color of silver chromate remains permanent on stirring. The proportion of chlorin is then calculated from the number of c. c. of silver solution added. For greater accuracy a second determination may be made, using as a comparison the liquid first titrated, the red color having been previously discharged by a few drops of sodium chlorid solution.

The water should always be as nearly neutral as possible before titration. If acid, it may be neutralized by the addition of sodium carbonate.

The residue obtained by evaporating the water with sodium carbonate as described in connection with the determination of the total solids will serve conveniently for estimating the chlorin. It is best to use 200 c. c. of the sample and redissolve the residue in about 50 c. c. of distilled water, rubbing the sides of the dish well with a rubber-tipped rod, and then titrating as indicated above.

NITROGEN IN AMMONIUM COMPOUNDS AND IN ORGANIC MATTER.

The nitrogen in ammonium compounds, and a part of that in the organic matter, is determined by a process of distillation first developed fully by Messrs. Wanklyn, Chapman, and Smith. It depends upon the conversion of the nitrogen into ammonium compounds and subsequent estimation in the distillate.

Apparatus Required :—

Distilling Apparatus.—That shown in Fig. 3 has been found to be the most convenient. The still consists of a

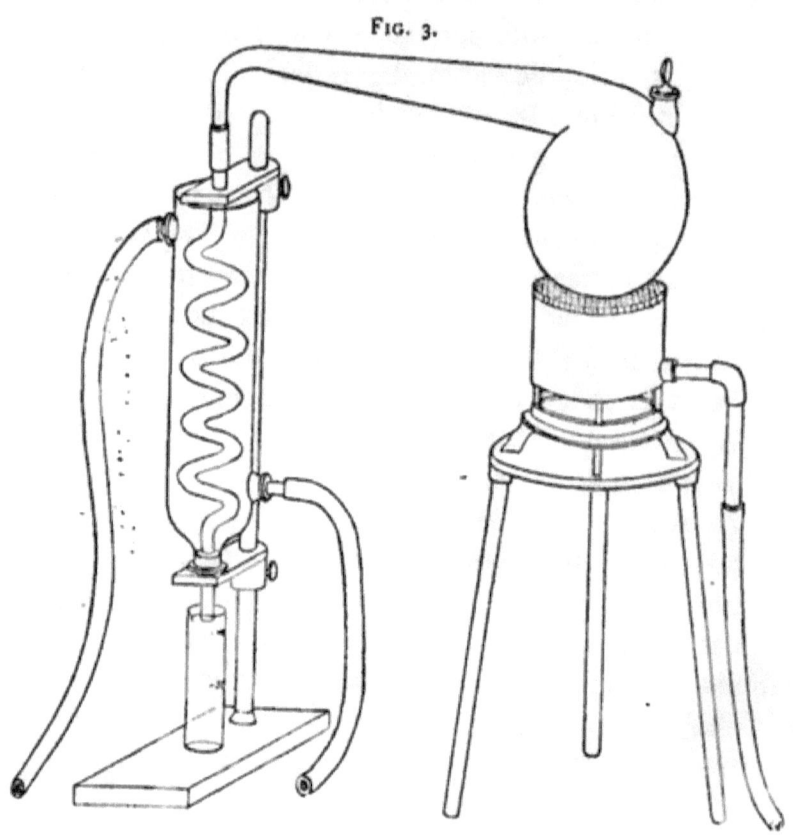

FIG. 3.

Bohemian glass retort of about 1000 c. c. capacity. The beak of the retort should incline slightly upward, to prevent contamination by splashing. At about two inches from the end it should be bent at a right angle, and drawn out so as to enter the condensing worm for about an inch, and terminate beneath the level of the water. Glass worms are apt to crack, and it is more satisfactory to use

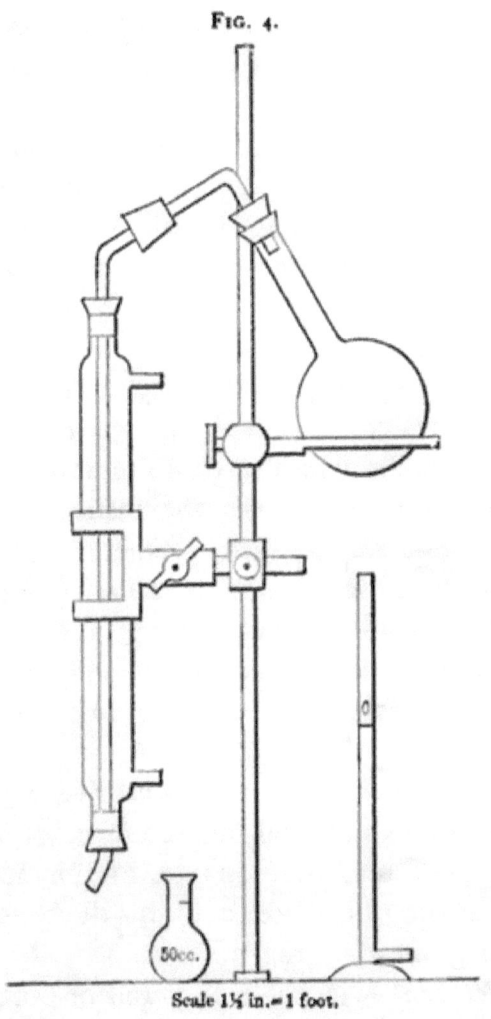

Fig. 4.

Scale 1½ in.= 1 foot.

one of block tin. A piece of rubber tubing is drawn over the junction. A rapid current of cold water should be maintained through the condenser. The heat is applied by means of the low-temperature burner, the iron ring of which is removed so that the retort rests directly on the gauze. With this arrangement the heat is under perfect control, and the danger of fracturing the glass is reduced to a minimum. It is advisable to protect the retort from drafts of cold air.

Another convenient form of apparatus is shown in Fig. 4. It is employed in the laboratory of the Massachusetts State Board of Health. (A. H. Gill, *The Analyst*, XVIII.)

Cylinders for Comparison-Color Tests, about 2.5 cm. in diameter and holding 100 c. c., made of colorless glass.

Solutions Required :—

Sodium Carbonate.—50 grams of pure sodium carbonate are strongly heated, dissolved in 250 c. c. of distilled water, and the solution boiled down to 200 c. c.

Ammonium-Free Water.—If the distilled water of the laboratory gives a reaction with Nessler reagent, it should be treated with sodium carbonate, about one grain to the liter, and boiled until about one-fourth has been evaporated. Ammonium-free water may be obtained by distilling, in a retort, water made slightly acid with sulphuric acid.

Standard Ammonium Chlorid.—Dissolve 0.382 gram of pure dry ammonium chlorid in 100 c. c. of ammonium-free water. For use, dilute 1 c. c. of this solution with pure water to 100 c. c. 1 c. c. of this dilute solution contains .00001 gram of nitrogen.

Nessler Reagent.—Dissolve 35 grams of potassium iodid in 100 c. c. of water. Dissolve 17 parts of mercuric chlorid in 300 c. c. of water. The liquids may be heated to aid solution, but must be cooled before use. Add the mercuric chlorid solution to that of the potassium iodid, until a permanent precipitate is produced. Then dilute with a 20 per cent. solution of sodium hydroxid to 1000 c. c., add mercuric chlorid solution until a permanent precipitate again forms and allow to stand until clear. Nessler and other reagents are best kept in glass-capped bottles, Fig. 5, in which the pipette may remain when not in use. The solution improves by keeping.

Alkaline Potassium Permanganate.—Dissolve 200 grams of potassium hydroxid, in sticks, and eight grams of potassium permanganate, in a liter of distilled water.

FIG. 5.

The solution may be boiled until about one-fourth is evaporated and then made up to a liter with ammonium-free water. It will still furnish some ammonium. Fox recommends to distill 50 c. c. with 500 c. c. of absolutely ammonium-free water, best twice distilled with sulphuric acid, and note the ammonia obtained. This quantity should be deducted in each analysis. The method of determining nitrogen by permanganate, as recommended by the A. A. A. S., avoids the necessity for this preliminary valuation of the solution.

Analytic Process:—

The retort and condenser are thoroughly rinsed with ammonium-free water, 500 c. c. of the water to be tested introduced, about five c. c. of the sodium carbonate solution added to render the water alkaline, and a piece of pumice-stone heated to redness and dropped in while hot. The water is then boiled gently until the distillate measures 50 c. c. The distillate is transferred to one of the color-comparison cylinders and two c. c. of Nessler reagent added. A yellowish-brown color is produced, the

c

intensity of which is proportional to the amount of ammonium present. The full color is developed in five minutes. This color is exactly matched by introducing into another cylinder 50 c. c. of ammonium-free water, some of the standard ammonium chlorid solution, and two c. c. Nessler reagent, as before. According as the color so produced is deeper or lighter than that obtained from the water, other comparison liquids are prepared containing smaller or larger proportions of the ammonium chlorid, until the proper color is produced.

The distillation is continued, successive portions of 50 c. c. each collected, and tested until no reaction occurs with Nessler reagent. The sum of the figures from the several distillates gives the total nitrogen obtainable as "free ammonia," so-called.

If the quantity of ammonium is sufficient to cause a precipitate, the color comparison cannot be accurately made. In most cases this will not be of serious moment, as the quantity will be beyond the allowable limit. If accurate determination be desired, it may be made by dividing the first distillate into two equal parts, nesslerizing one of these, and then, if necessary, diluting the second part with ammonium-free water and nesslerizing this.

Occasionally the evolution of ammonium hydroxid continues indefinitely, and may even increase with successive distillates. This is due, not to ammonium compounds existing as such, but to decomposition of certain nitrogenous bodies, especially urea. In this case, it is not advisable to prolong distillation beyond the fourth or fifth distillate, but to proceed to the following part of the process.

The residue in the retort serves for the determination of the nitrogen which is convertible into ammonium by alkaline potassium permanganate—the so-called "albuminoid ammonia" of Messrs. Wanklyn, Chapman, and Smith.

Fifty c. c. of alkaline permanganate solution are added to the retort, the distillation resumed, and the nitrogen estimated in each 50 c. c. as before, deducting that yielded by the permanganate.

It is the practice of some analysts to mix the distillates of each of the above operations, and thus make determinations merely of the total nitrogen in each case. By so doing valuable information may be lost, since it has been pointed out by several observers, notably Mallet and Smart, that important information may be gained by observing the rate of evolution of the ammonium hydroxid. Mallet has further pointed out that many waters may contain substitution ammoniums which may pass over before the addition of the alkaline permanganate, but not be correctly measured by nesslerizing. To avoid this source of error, he suggested that two determinations be made on each sample, one as above described and the other by the addition of alkaline permanganate without previous distillation. In this manner a higher figure will often be obtained than the sum of the figures from the two distillations by the other process.

Since small quantities of ammonium compounds and nitrogenous matters are everywhere present, the greatest care should be exercised in order to avoid their introduction in any way during the course of the analysis. All measuring vessels, cylinders, etc., should be thoroughly rinsed before using. The temperatures of the distillates and standards should be approximately the same when the colors are compared.

The Chemical Section of the American Association for the Advancement of Science recommends the following method for the application of the process embodying the results of recent investigations:—

"Two hundred c. c. of distilled water, together with 10 c. c. of the sodium carbonate solution, are distilled down to about 100 c. c. in the retort in which the analysis is to be conducted, and the last portion of 50 c. c. nesslerized to assure freedom from ammonium. Then 500 c. c. of the water to be examined are added and the distillation is carried on at such a rate that about 50 c. c. are collected in each succeeding ten minutes, and until a 50 c. c. measure of distillate is obtained containing only an inappreciable quantity of ammonia. In nesslerizing, five minutes are to be allowed for the full development of color; after this, no change takes place for many hours.

"Now throw out the contents of the retort, rinse it thoroughly, put in 200 c. c. of distilled water and 50 c. c. of the permanganate solution, distill down to about 100 c. c., and nesslerize the last portion of 50 c. c., to make sure of freedom from ammonia; add another portion of 500 c. c. of the water under examination and proceed with the distillation and nesslerizing as with the first portion.

"The difference between the 'free' ammonia of the first operation and the total ammonia of the second, is to be taken as the 'albuminoid' ammonia."

Bachman has described (*Jour. of Amer. Chem. Soc.*, April, 1895) a process for determining nitrogen by permanganate by using an apparatus which permits of the addition of the water in small and limited quantity to the full strength of permanganate solution at about the rate of distillation, so that at no time does it act upon a large volume. The apparatus consists of a flask of one liter capacity, fitted by a ground joint to a hollow cap carrying two upright tubes, each provided with a stop-cock, and to which cap also is attached the tube through which the distilled vapors pass to the condenser. Of the two upright tubes, one has a capacity of 50 c. c., and its end reaches through the cap to within three inches of the bottom of the distillation flask. The other has a capacity of 250 c. c., and its tube projects to within one-half of the bottom. The condensing apparatus is a closely-coiled glass worm, with a long projection at the lower extremity which passes through a soft rubber cork, inserted in a receiver. Through this cork also there is connected a Will and Varrentrap ammonium-absorption bulb. All connections should be tight without the use of a lubricant. The arrangement can be made movable so as to be raised from the bath if necessary. Bachman employs a bath of salt water, but it is not unlikely that a low temperature burner would answer every purpose.

The procedure is as follows: After thoroughly rinsing the apparatus with ammonium-free distilled water, 500 c. c. of the water under examination with the addition of a little sodium carbonate if needed, are put into the flask. The Will and Varrentrap bulb is charged with Nessler solution, and the free end of the bulb connected with an exhaustion apparatus. A partial vacuum is established gradually, and the distillation carried on until 200 c. c. have passed over. The vacuum-tube is disengaged and 250 c. c. of the liquid in the distilling flask drawn into the larger upright tube, and 50 c. c. of alkaline permanganate solution which has been previously placed in the small tube, allowed to run in, after the flask to receive the distillate has been again attached. This arrangement gives 50 c. c. of alkaline permanganate acting on 50 c. c. of water, and after 30 or 40 c. c. have been distilled over the water in the larger tube is allowed to drop in at the same rate as the distillation, which must not exceed 50 c. c. in 15 minutes. This is continued until from 250 to 300 c. c. have been distilled over, which is then nesslerized.

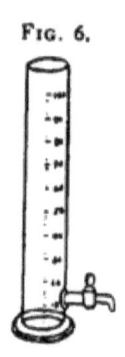

Fig. 6.

For nesslerizing and other color comparisons, many forms of apparatus have been proposed. One of the simplest is that devised by Hehner, shown in Fig. 6. It consists of a graduated cylinder with a stopcock near the base, by which the liquid can be drawn down at will. Two such cylinders may be used, one for the nesslerized distillate, the other for the comparison liquid. The darker liquid is drawn out until the tints are equal, when the relative volumes remaining will give the data for calculation.

TOTAL ORGANIC NITROGEN.

Several processes for the determination of the organic nitrogen in water, based on those in use in ordinary organic analysis, have been devised. That of Frankland and Armstrong requires complex and extensive apparatus and special skill, has been shown also to be liable to inaccuracies, and has not come into extended use.

The ease and certainty with which the nitrogen of most organic bodies may be converted into ammonium sulphate by boiling with sulphuric acid, offers a means of determination free from the objections of former methods. The method introduced by Kjeldahl for general organic analysis, was first successfully applied to water analysis by Drown and Martin (*Technology Quarterly*, II, 3).

In their original process 500 c. c. was concentrated to about 300 c. c., and the distillate nesslerized for determining the nitrogen existing as ammonium compounds. The organic nitrogen is then determined in the residual water. Owing to the fact that in many waters the organic matter is decomposed by boiling, there is liability to underestimation of the nitrogen. It is best, therefore, to determine at once the total unoxidized nitrogen, and estimate, without distillation, on a separate portion of the sample, the nitrogen that exists in ammonium compounds. The procedure is as follows:—

Reagents Required:—

Concentrated Sulphuric Acid.—This should be as free as possible from nitrogen. It can be obtained containing only 0.015 mgm. in 10 c. c.

Sodium Hydroxid Solution.—The white granulated caustic soda sold for household use will answer; 350 grams are dissolved in water and made up to 1000 c. c.

Sodium Carbonate and Hydroxid Solution.—25 grams of each are dissolved in 250 c. c. of distilled water, and the solution boiled down to 200 c. c., to free it from ammonium.

Analytic Process:—

Determination of Nitrogen Existing as Ammonium.—200 c. c. of the water are placed in a stoppered bottle, two c. c. each of the solutions of sodium carbonate and sodium hydroxid added, the stopper inserted, the solutions mixed and allowed to stand for an hour or two. A filter is prepared by inserting a rather large plug of absorbent cotton in a funnel. This should be washed with ammonium-free water until the filtrate gives no color with Nessler reagent. The clear portion of the sample is drawn off with a pipette and run through the filter, the first portions being rejected, since it is diluted by the water retained in the cotton. The filtration is rapid, and when 100 c. c. of the liquid have passed through it is nesslerized. If but little ammonium is present, a narrow tube about 60 centimeters long should be used for observing the color.

Estimation of the Total Organic and Ammoniacal Nitrogen.—500 c. c. of the water are placed in a round-bottomed Bohemian glass flask, 10 c. c. of concentrated sulphuric acid added, and a piece of pumice-stone is heated to bright redness and dropped in while hot. The liquid is boiled until the acid is colorless or very pale, the boiling being continued for nearly an hour from this point. The flask is then removed from the flame, allowed to cool, and about 250 c. c. of ammonium-free water added. 50 c. c. of the sodium hydroxid solution should be placed in the distilling apparatus, Fig. 3, about 250 c. c. of water added, a piece of red-hot pumice-stone dropped in and the liquid distilled until the distillate is free from ammonium. It is best to distill until the retort contains not more than 100 c. c. The sulphuric acid solution is then poured in slowly, by means of a funnel, the stem of which touches the side of the retort, so that the two liquids do not mingle. The stopper of the retort is inserted, the liquids mixed by gentle agitation, and distilled. If much ammonium is present it is advisable to distill the first portion into about 10 c. c. of very dilute (1 : 1000) sulphuric acid, a piece of glass tube being connected to the condensing worm so that the lower end dips below the surface of the liquid. The distillates are collected and nesslerized in the usual way.

A blank experiment should be made to determine the amount of ammonium in the sulphuric acid.

NITROGEN AS NITRATES.

Solutions Required :—

A. H. Gill (*Jour. of Amer. Chem. Soc.*, 1894), has subjected the various indirect methods of estimating nitrates to comparative examination, and prefers the following reagents :—

Phenoldisulphonic Acid.—Strong sulphuric acid and pure phenol are mixed in the proportion of 37 grams of the former to three grams of the latter, and heated for six hours *in*, *not upon*, the water-bath. The resulting compound usually solidifies to a white mass on standing, but can be easily liquified on the water-bath during the evaporation of the samples to be tested.

Standard Potassium Nitrate.—0.722 gram of potassium nitrate, previously heated to a temperature just sufficient to fuse it, are dissolved in water, and the solution made up to 1000 c. c. One c. c. of this solution will contain .0001 grm. of nitrogen.

Analytic Process :—

A measured volume of the water is evaporated just to dryness in a porcelain basin about six cm. in diameter. One c. c. of the phenoldisulphonic acid is added and thoroughly mixed with the residue by means of a glass rod. The liquid is then diluted with about 25 c. c. of water, ammonium hydroxid added in excess, and the solution made up to 50 c. c.

The nitrate converts the phenoldisulphonic acid into

picric acid, which by the action of the ammonium hydroxid forms ammonium picrate; this imparts to the solution a yellow color, the intensity of which is proportional to the amount present.

One c. c. of the standard solution of potassium nitrate is now similarly evaporated in a platinum basin, treated as above, and made up to 50 c. c. The color produced is compared to that given by the water, and one or the other of the solutions is diluted until the tints of the two agree. The comparative volumes of the liquids furnish the necessary data for determining the amount of nitrate.

Mr. Gill has furnished me with an account of the method pursued at the Massachusetts Institute of Technology for obtaining a comparison liquid. Ten c. c. of the standard nitrate solution is cautiously evaporated, spontaneously or in a desiccator, and three c.c. of the phenoldisulphonic acid added, the mixture diluted with water to exactly 1000 c. c. Of this solution, portions are measured out with a finely and accurately graduated pipette into the porcelain dish similar to that in which the water-sample has been evaporated, water added to make exactly 10 c. c. and then two or three drops of ammonium hydroxid to make the liquid alkaline. The quantity of liquid taken in the pipette has the following values in parts per million of nitrogen as nitrates: 10 c. c. $= 1.0$; 5 c. c. $= 0.5$; 2 c. c. $= 0.2$; 0 5 c. c. $= 0.05$. Comparison is thus made directly in the dishes. In case of doubt the liquids are compared in tubes similar to nesslerizing tubes, but cut down to hold 15 or 20 c. c.

The results obtained by this method are satisfactory. Care should be taken that the same quantities of phenoldisulphonic acid are used for the water and for the comparison liquid.

With subsoil and other waters probably containing much nitrates, 10 c. c. will be sufficient; but with river and spring waters, 25 c. c. may be used. When the organic matter is sufficient to color the residue, it will be well to purify the water by addition of aluminum hydroxid and filtration, before evaporating.

Chlorin interferes with the accuracy of the test, but Gill finds that when not amounting to more than 20 parts per million it does not impair the practical value of the results. When greater than this, it is best to evaporate in vacuo over sulphuric acid. If the chlorin be more than 70 parts per million it should be considerably reduced by the addition of silver sulphate, which has been ascertained to be free from nitrates. Nitrites do not influence the reaction.

The following is the process for determining nitrogen as nitrates (and nitrites) recommended by the Chemical Section of the A. A. A. S. It depends upon conversion into ammonium by the copper-zinc couple, and subsequent nesslerizing. It is inferior to the phenoldisulphonic acid method, both in convenience and accuracy, and does not exclude the influence of nitrites.

Take two wide-mouth glass-stoppered bottles, each holding 250 c. c. and a piece of sheet zinc as long and about as wide as the bottles are deep from the shoulder down; clean the zinc by dipping in dilute acid and washing with water and make it into a loose coil by rolling it round a piece of glass tube. Immerse it in a 1.4 to 1.8 per cent. solution of cupric sulphate in ammonia-free water, and leave it there until its surface is well covered with a continuous layer of the black copper; lift it out carefully, cover it in a beaker with successive portions of ammonia-free water, lifting it out and draining each time, and at once put it into one of the bottles of acidified water, prepared as follows:—

Make 500 c. c. of the water to be examined distinctly acid with oxalic acid added in fine powder, with constant stirring, so that it shall dissolve readily, and pour half of the liquid into one of these 250 c. c. bottles, and half into the other, and leave them, stoppered, in a warm place for twenty-four hours. Then nesslerize both samples, decanting off the portions as wanted, from the precipitated earthy oxalates, and using double the usual quantity of Nessler's solution, since the free oxalic acid has to be neutralized first by the alkali of the reagent. The proportion of ammonia may often be so large in the water in which the reduction is made, by the copper-zinc couple, that only five or ten c. c. can be taken for each test, and made up to 50 c. c. by the addition of ammonia-free water. The difference between the results with the two portions of water, gives the amount of nitrogen due to the oxidized nitrogen compounds in the water examined.

NITROGEN AS NITRITES.

The following is Ilosvay's modification of Griess's test. It has the advantage over the original method, that the color is developed more rapidly, and the solutions are not so liable to change.

Solutions Required :—

Paramidobenzenesulphonic Acid Solution (Sulphanilic Acid).—Dissolve 0.5 gram in 150 c. c. of diluted acetic acid, sp. gr. 1.04.

a-amidonaphthalene Acetate Solution.—Boil 0.1 gram of solid a-amidonaphthalene (naphthylamin) in 20 c. c. of water, filter the solution through a plug of washed absorbent cotton, and mix the filtrate with 180 c. c. of diluted acetic acid. All water used must be free from nitrites, and all vessels must be rinsed out with such water before tests are applied, since appreciable quantities of nitrites may be taken up from the air.

Standard Sodium Nitrite.—0.275 gram pure silver nitrite are dissolved in pure water, and a dilute solution of pure sodium chlorid added until the precipitate ceases to form. It is then diluted with pure water to 250 c. c., and allowed to stand until clear. For use 10 c. c. of this solution are diluted to 100 c. c. It is to be kept in the dark.

One c. c. of the dilute solution is equivalent to .00001 gram nitrogen.

The silver nitrite is prepared thus: A hot concentrated solution of silver nitrate is added to a concentrated solution of the purest sodium or potassium nitrite available, filtered while hot and allowed to cool. The silver nitrite will separate in fine needle-like crystals, which are freed from the mother liquor by filtration by the aid of a filter pump. The crystals are dissolved in the smallest possible quantity of hot water, allowed to cool and again separated by means of the pump. They are then thoroughly dried in the water-bath, and preserved in a tightly-stoppered bottle away from the light. The purity may be tested by heating a weighed quantity to redness in a tared porcelain crucible and noting the weight of the metallic silver. 154 parts should leave a residue of 108 parts silver.

Analytic Process:—

25 c. c. of the water are placed in one of the color-comparison cylinders, and 2 c. c. each of the test solutions are dropped in. It is convenient to have a pipette for each solution, and to use it for no other purpose.

One c. c. of the standard nitrite solution is placed in another clean cylinder, made up with nitrite-free water to 25 c. c. and treated with the reagents as above.

In the presence of nitrites a pink color is produced. At the end of five minutes the two solutions are compared, the colors equalized by diluting the darker, and the calculation made as explained under the estimation of nitrates.

The reactions consist in the conversion of the sulphanilic acid into diazobenzenesulphonic anhydrid, by the nitrite present; this compound is then in turn converted by the amidonaphthalene into azo-α-amidonaphthalene-parazobenzene sulphonic acid. The last-named body gives the color to the liquid.

OXYGEN-CONSUMING POWER.

All organic materials being more or less easily oxidized, several methods have been suggested for determining the oxygen-consuming powers of waters by treatment with active oxidizing agents. These methods are, however, limited in value. The organic matters in water differ much in character and condition, and their oxidability is subject to much variation, according to the circumstances under which the test is made. Nevertheless, as a high oxygen-consuming power certainly indicates departure from purity, some additional evidence may be obtained. Potassium permanganate is especially suitable. The test is usually made by introducing a known amount of the permanganate into the water, which has been rendered acid, and measuring after a definite period the proportion which has been decomposed.

It must not be overlooked that if a water contains nitrites, ferrous compounds, or sulphur compounds other than sulphates, the proportion of oxygen consumed will be greater than that required for the organic matter. It has been proposed, in order to remove the nitrites before applying the permanganate, to take 500 c. c. of the water, add 10 c. c. of the dilute sulphuric acid, boil for twenty minutes, allow to cool, and then treat with permanganate. Since, however, the amount of nitrites, if appreciable, can be directly determined, it is more satisfactory to deduct from the oxygen consumed the amount required to convert the nitrites present into nitrates, and the remainder will be that required for the other oxidizable ingredients. 14 parts of nitrogen existing as nitrite require 16 parts of oxygen for conversion into nitrate. Similarly, 112 parts of iron in a ferrous compound will require 16 parts of oxygen for conversion to the ferric condition.

Of the following methods the first is due in the main to Dr. Tidy, has been improved by Dr. Dupré, and is approved by the Society of Public Analysts of Great Britain :—

Solutions Required :—

Standard Permanganate.—.395 gram pure potassium permanganate are dissolved in distilled water, and the solution made up to 1000 c. c. One c. c. is equal to .0001 gram oxygen.

Diluted Sulphuric Acid.—Add 50 c. c. of pure sulphuric acid to 100 c. c. of water, and then add solution of potassium permanganate until a faint pink color is obtained, which is permanent when the liquid is heated to 80° Fahrenheit for four hours.

Potassium Iodid.—10 grams of the pure salt recrystallized from alcohol are dissolved in 100 c. c. of distilled water.

Sodium Thiosulphate.—One gram of the pure crystallized salt dissolved in 2000 c. c. of distilled water.

Starch Indicator.—One gram of clean starch is mixed smoothly with cold water into a thin paste, then poured gradually into about 200 c c. of boiling water, the boiling continued for one minute, the liquid allowed to settle, and the clear portion used. It is best freshly prepared.

Analytic Process:—

Two determinations are made, one, of the oxygen consumed in fifteen minutes, which is considered to represent the nitrites, sulphids, or ferrous compounds, and the other of the oxygen consumed by four hours' action. Both determinations are made at a temperature of 80° F. Three glass stoppered bottles, of about 350 c. c. capacity, are rinsed with strong sulphuric acid, and then with water. In one is placed 250 c. c. of pure distilled water as a control experiment, and in each of the others 250 c. c. of the water to be tested. The bottles are stoppered and brought to a temperature of 80° F.; 10 c. c. of the dilute sulphuric acid and 10 c. c. of the standard permanganate are added to each and the stoppers again replaced. At the end of fifteen minutes one sample of water is removed from the bath, and two or three drops of the potassium iodid solution added to remove the pink color. After thorough admixture, the thiosulphate solution is run in from a burette until the yellow color is nearly destroyed, a few drops of the starch solution added, and the addition of the thiosulphate continued until the blue color is quite discharged. If the addition of the thiosulphate solution has been properly conducted, one drop of the permanganate solution will restore the blue color.

The other bottles are maintained at 80°F. for four hours. Should the pink color disappear rapidly in the bottle containing the water under examination, 10 c. c. of the permanganate solution must be added to each bottle, in order to maintain a distinct pink color. At the end of four hours each bottle is removed from the bath, two or three drops of potassium iodid added, and the titration with thiosulphate solution conducted as just described. The calculation is most conveniently made as follows:—

$a =$ number of c. c. required for the control experiment.
$b =$ number of c. c. required for the water under examination.
$c =$ available O in permanganate (.001 for 10 c. c.).
$x =$ oxygen consumed by water.
Then, $a : a-b :: c : x$.

The following method is recommended by the Chemical Section of the American Association for the Advancement of Science:—

"Prepare a solution of potassium permanganate containing 0.1 mgm. of available oxygen to one c. c. and a solution of oxalic acid of such strength as to decompose the permanganate solution, volume for volume, the strength being re-determined from time to time. The water used for making these solutions should be purified by distillation from alkaline permanganate.

"To 200 c. c. of water to be examined, in a 400 c. c. flask, add 10 c. c. of dilute sulphuric acid (1 : 3) and such measured quantity of the permanganate as will give a persistent color; boil ten minutes, add, if necessary, more permanganate in measured quantities so as to maintain the red color; remove the flask from the lamp, add 10 c. c. of oxalic acid solution to destroy the color, or more if required by the excess of permanganate, and then add permanganate, drop by drop, till a faint pink tint appears.

From the total quantity of permanganate used deduct the equivalent of the oxalic acid used, and from the remainder calculate the milligrams of oxygen consumed by the oxidizable organic matter in the water."

The oxygen-consuming power may also be indirectly estimated by the action of the organic matter upon silver compounds. H. Fleck's method (*Fresenius' Quantitative Analysis*, English edition), depends upon the reduction produced by boiling the water with alkaline solution of silver thiosulphate and estimation of the unreduced silver. A. R. Leeds (*Lond., Edin. and Dub. Phil. Mag.*, July, 1883) gives a method by treating the water with decinormal silver nitrate, exposing to light until it settles perfectly clear, and estimating the reduced silver in the deposit.

These methods are open to practically the same objections as in the use of permanganate, and do not seem to possess any decided advantage. Qualitative results of some interest may occasionally be obtained by the following method: Two c.c. of a one per cent. solution of silver nitrate, rendered decidedly alkaline by ammonium hydroxid, are added to 100 c. c. of the water in a stoppered bottle, which is then placed in full sunlight for two hours. Waters containing but little organic matter will not show at the end of this period any appreciable tint. The following results will show the character of the test:—

    Schuylkill water, ... no color.
       "   " with 0.02 c. c. urine, . . red-brown.
       "   " with 0.01 c. c. urine, . . deep brown.
       "   " with 0.25 gram raw sugar. no color.
    Well water, not perfectly pure, but not unfit to drink, ... faint black.
       "   " markedly contaminated, . . black ppt. almost immediately.
    Water from a small stream quite pure, . . no color.

PHOSPHATES.

Solutions Required:—

Ammonium Molybdate.—Ten grams of molybdic anhydrid are dissolved in 41.7 c. c. of ammonium hydroxid, sp. gr. 0.96, and the solution is poured slowly, and with constant stirring, into 125 c. c. of nitric acid, sp. gr. 1.20, and allowed to stand in a warm place for several days, until clear.

Analytic Process:—

500 c. c. of the water are slightly acidified with nitric acid, and evaporated to about 50 c. c. A few drops of dilute solution of ferric chlorid are added and then ammonium hydroxid in slight excess. The precipitate, which contains all the phosphate, is filtered off and dissolved on the filter by the smallest possible quantity of hot dilute nitric acid. The filtrate and washings should not exceed five c. c.; if more, they must be evaporated to this bulk. The liquid is heated nearly to boiling, two c. c. of ammonium molybdate solution added, and the liquid kept moderately warm for half an hour. If the quantity of precipitate is appreciable, it is collected on a small weighed filter, washed with distilled water, dried at 100° F. and weighed. The weight of the precipitate multiplied by 0.05 gives the amount of PO_4. If the quantity is not sufficient to collect in this manner, it is usually reported, according to circumstances, as "traces," "heavy traces," or "very heavy traces."

DISSOLVED OXYGEN.

The method here given, a modification of Mohr's, was proposed by Blarez. It is rapid and satisfactory.

Solutions Required:—

Sodium Hydroxid.—40 grams of pure sodium hydroxid to the liter.

Ferrous-Ammonium Sulphate.—40 grams dissolved in about a liter of water, and acidified with a few drops of concentrated sulphuric acid.

Decinormal Potassium Permanganate.—3.156 grams dissolved in a liter of distilled water. The accuracy of this solution should be determined by titration with a known weight of ferrous-ammonium sulphate. One c. c. should be equivalent to .0392 grm. ferrous-ammonium sulphate (.0008 gram of oxygen).

The apparatus employed (shown in Fig. 7) is a globular separator, of about 250 c. c. capacity. Above the bulb is a caoutchouc stopper carrying a cylindrical funnel, of about 12 c. c. capacity, terminating in a tube, 8 mm. calibre, sharply contracted at the outlet to a capillary opening. The tube should project about 6 mm. below the stopper. The exact capacity of the apparatus is measured as follows: The bulb is completely filled with water and the stopper inserted; the level of the water will rise slightly in the funnel tube, and should be brought down to its outlet by drawing a little off at the stopcock, after which the water is run into a graduated measure and its volume noted.

Fig. 7.

Analytic Process:—

35 c. c. of mercury and 10 c. c. of sodium hydroxid solution are put into the bulb, and then sufficient of the water to be tested to fill it. The funnel stopper is inserted and the water which rises into the funnel brought into the bulb by

cautiously running out at the stopcock, mercury, the volume of which should be noted. The exact volume of water used is thus known. Five c. c. of the ferrous-ammonium sulphate solution are poured into the funnel, brought into the bulb by running out mercury, and the liquid thoroughly mixed by giving the apparatus a gyratory movement. After standing five or six minutes the oxygen will be completely absorbed; 10 c. c. of the diluted sulphuric acid are now added by the same method. On agitating the bulb the contents become clear. The watery liquid is then transferred to a beaker and titrated with decinormal permanganate. A volume of water equal to that used in the test is poured into another beaker, 10 c. c. each of the sodium hydroxid and diluted sulphuric acid added, and then five c. c. of ferrous-ammonium sulphate solution. The resulting liquid is titrated with permanganate. The weight of oxygen corresponding to the difference between the two titrations gives the weight of dissolved oxygen in the liquid employed. From this should be subtracted as correction the amount of oxygen dissolved by a volume of water equal to that of the sodium hydroxid solution used. This is found by reference to the table in the appendix. The amount of oxygen dissolved in the sulphuric acid has no appreciable effect.

Nitrates do not appear to impair the accuracy of this method, and the interfering action of nitrites and other reducing compounds is avoided by the control experiment as detailed.

It is perhaps hardly necessary to add that the exact temperature of the water is to be noted at the time of collection of the sample.

In transferring to the bulb, the water should be agitated

as little as possible in contact with the air, in order to avoid the absorption of oxygen. A siphon should be used for this purpose, the lower end being allowed to reach to the bottom of the bulb.

The following modification is suggested as being especially suitable for poorly oxygenated waters: An accurately stoppered bottle, the exact capacity of which is known (about 500 c. c. is a convenient size), is completely filled at the source with the water to be examined, and the stopper inserted so as to drive out all air. The stopper is removed in the laboratory, 50 c. c. of the water drawn off with a pipette, and the water covered immediately with a layer of gasoline previously purified by shaking up several times with a solution of potassium permanganate and diluted sulphuric acid, and washed several times with water. The sodium hydroxid, ferrous-ammonium sulphate, and sulphuric acid are introduced into the water by means of burettes to which long glass delivery tubes are attached. The titration with potassium permanganate is conducted in the same way. The liquid is mixed from time to time, as the solutions are added, by means of a glass rod. In this way the air may be completely excluded throughout the entire operation. The amount of water titrated is, of course, equal to the whole capacity of the bottle, less the 50 c. c. removed by the pipette.

The control experiment on an equal volume of the water, and the correction for the oxygen added with the sodium hydroxid solution, are made as detailed above.

Dupré has employed the determination of free oxygen for the estimation of the proportion of oxygen-consuming microbes. The principle of the method is that pure water, if kept in a closed bottle, will neither gain nor lose oxygen in any length of time, but if organisms capable of causing absorption of oxygen are present, the quantity will decrease.

The experiment is carried out by placing a sample of the water in a clean bottle, and vigorously shaking it to saturate with air. A clean 250 c. c. bottle is completely filled with the water, tightly stoppered, and maintained at a temperature of 68° F. for ten days; the free oxygen remaining is then determined.

POISONOUS METALS.

Under this conventional title are included *barium, chromium, zinc, arsenum, copper*, and *lead; manganese, iron, aluminum* also, though not usually classed in this group, are objectionable when present in notable amounts.

Barium is rarely present, and only in water containing no sulphates. It can be detected and estimated by slightly acidifying the water with hydrochloric acid, filtering if necessary, and adding solution of calcium sulphate. The precipitated barium sulphate is collected and weighed in the usual way.

Chromium is rarely present, but may be looked for in the waste waters of dye works and similar sources. To detect it, a considerable volume of the water is evaporated to dryness with addition of a small amount of potassium chlorate and nitrate, transferred to a porcelain crucible and brought to quiet fusion; any chromium present will be found in the residue in the form of chromate. The fused mass, after cooling, is boiled with a little water, filtered, the filtrate rendered slightly acid with hydrochloric acid, and a solution of hydrogen dioxid added. In the presence of chromium a transient blue color will appear; by adding a little ether, and shaking the mixture the color will pass into the ether, and on standing will form a blue layer on the surface of the water.

Zinc is best detected by the test described by Allen. The water is rendered slightly alkaline by addition of ammonium hydroxid, heated to boiling, filtered, and the clear liquid treated with a few drops of potassium ferrocyanid; in the presence even of the merest trace of zinc a white precipitate will be produced.

Arsenum is most readily detected by Reinsch's test. One liter of the water is rendered slightly alkaline by pure sodium carbonate, and evaporated nearly to dryness in a porcelain basin. Two or three c. c. of water strongly acidulated with hydrochloric acid are placed in a small test-tube, about ½ square centimeter of bright copper foil is added, and the liquid boiled gently for a few moments. If the copper remains bright, showing that the reagents contain no arsenum, the water-residue is acidified with hydrochloric acid, added to the contents of the test-tube, and the liquid again boiled for several minutes. If arsenum be present, a steel-gray stain will appear on the copper. The slip is removed, washed with distilled water, *thoroughly* dried by pressure between filter paper, inserted into a narrow glass tube closed at one end, which has been previously dried by heating nearly to redness. The tube is gently heated at the point at which the copper rests; the deposit will sublime and collect on the cooler portion of the tube, in crystals which the microscope shows to be octahedral.

Since small amounts of arsenum frequently occur in reagents and in glass vessels, care must be taken to avoid such sources of error. Sodium carbonate solution may contain arsenum dissolved from the glass bottle in which it is kept. It is best, therefore, to use the solid carbonate for rendering the water alkaline, and to determine its purity before use.

Iron is detected by the addition of a drop of ammonium sulphid to the water in a tall glass cylinder. Ferrous sulphid is formed, having a greenish-black color, instantly discharged by acidifying the water with dilute hydrochloric acid. A still better test is the production of a blood-red color, with potassium thiocyanate, due to the formation of ferric thiocyanate. The water should be first boiled with a few drops of nitric acid, to convert the iron to the ferric condition, cooled, and a drop or two of the solution of potassium thiocyanate added. The test is very delicate. Either of the above tests may be made quantitative by matching the color produced in 100 c. c. of the water with that obtained from a known weight of iron. The method with potassium thiocyanate is preferable, as it is more delicate and there are fewer interfering conditions. The following is the method as elaborated by Thompson and described in Sutton's " Volumetric Analysis : "—

Solutions Required :—

Standard Ferric Sulphate.—0.7 gram ferrous ammonium sulphate are dissolved in water acidified with sulphuric acid, and potassium permanganate solution added until the solution turns a very faint pink color. The solution is diluted to a liter. One c. c. contains 0.1 milligram iron.

Diluted Nitric Acid.—30 c. c. concentrated nitric acid diluted with water to about 100 c. c.

Potassium Thiocyanate.—Five grams of the salt dissolved in about 100 c. c. water.

Analytic Process :—

About 100 c. c. of the water are evaporated to small bulk,

acidified with hydrochloric acid, and just sufficient dilute potassium permanganate solution added to convert all the iron to the ferric condition. The liquid is evaporated nearly to dryness to drive off excess of acid, then diluted to its original volume, 100 c. c. In two tall glasses marked at 100 c. c., five c. c. of the nitric acid and 15 c. c. of the thiocyanate solution are placed. To one of these a measured volume of the treated water is added and both vessels filled up to the mark with distilled water. If iron is present, a blood-red color will be produced. Standard iron solution is added to the second vessel until the color agrees. The amount of water which is added to the first glass will depend upon the quantity of iron it contains; not more should be used than will require two or three c. c. of the standard to match it, otherwise the color will be too deep for comparison.

Manganese.—The following method is described by Wanklyn in his treatise on water-analysis. About one liter of the water is evaporated to small bulk, nearly neutralized by hydrochloric acid and treated with a few drops of a solution of hydrogen dioxid. The formation of a brown precipitate indicates the presence of manganese. The test is very delicate. The precipitate may be collected on a filter, the filter ashed, and the residue fused with a mixture of sodium carbonate and potassium nitrate. Green potassium manganate will be produced, which, when boiled with water, will give a bright red solution of potassium permanganate. The quantitative determination is given elsewhere.

Aluminum.—Traces of this element are to be expected in all waters, and it is not usual to test for it except in elaborate analysis of the mineral ingredients, as described in another section. The use of aluminum sulphate as a coagulant in many rapid-filtration methods makes it necessary to examine effluents for excess of precipitant, and this may be done by the following method devised by Mrs. Richards :—

To 25 c. c. of the water to be tested (concentrated from one liter or more, if necessary) is added a few drops of freshly prepared logwood decoction ; any alkali is neutralized and the color is brightened by the addition of two or three drops of acetic acid. By comparison with standard solutions, the amount of alum present may be determined. One part of alum in 1,000,000 of water can be detected with certainty. In cases of greater dilution concentration of several liters may be necessary to obtain a decisive test. The logwood chips yield the right color *only* after having been treated with boiling water *two or three times, rejecting* the successive decoctions. The first portion gives a yellow color, the third or fourth usually a deep red.

Lead may be readily detected by adding to the water in a tall glass cylinder a drop of ammonium sulphid ; brownish black lead sulphid is formed, which does not dissolve either by acidulating the water with dilute hydrochloric acid (distinction from iron), nor by the addition of about one c. c. of a strong solution of potassium cyanid (distinction from copper). S. Harvey (*Analyst*, April, 1890) gives the following method for detecting lead in water: 250 c. c. are placed in a conical precipitating jar, about 0.1 gram of crystallized potassium dichromate is added and dissolved by agitation. The same volume of lead-free water is treated in the same manner, and the two solutions placed side by side. Water containing 0.3 parts per million, will show a turbidity in 15 minutes which will be rendered more distinct by contrast with the clear water alongside. By allowing the jar to stand for about twelve hours undisturbed, the precipitate will settle and will become still more distinct. No other metal likely to be present in water will give a similar reaction.

In the absence of copper the amount of lead present may be determined as follows: A solution is prepared containing 1.6 grams of lead nitrate to the liter; one c. c. of this contains one milligram lead. 100 c. c. of the water to be tested are placed in a tall glass vessel, made acid by the addition of a few drops of acetic acid and five c. c. of hydrogen sulphid added. In a similar vessel 100 c. c. of distilled water are placed, together with the same quantities of acetic acid and hydrogen sulphid, and sufficient of the standard lead solution to match the tint in the first cylinder. The amount of lead in the water under examination is thus known.

Copper is detected in the same manner as lead by acidifying the water with acetic acid and adding hydrogen sulphid. The precipitate is distinguished from lead sulphid by the fact that the color is discharged on the addition of about one c. c. of a strong solution of pure potassium cyanid. It may be further confirmed by the addition to another portion of the water of a solution of potassium ferrocyanid. In the presence of even a very small amount of copper, a mahogany red color is produced.

In the absence of lead, copper is estimated in the same way as that metal, using, however, a standard solution of copper for the comparison liquid. This is made by dissolving 3.929 grams of crystallized copper sulphate in one liter of water. One c. c. of the solution contains one milligram copper.

If both lead and copper are present, a large quantity of the water should be evaporated to small bulk, and the metals separated and estimated by any one of the ordinary laboratory methods.

The following table, prepared by A. J. Cooper, indicates the comparative delicacy of some of the ordinary tests for the detection of poisonous metals in water:—

Metal.	Reagent.	Depth of Liquid, $3\frac{1}{4}$ inches.	Depth of Liquid, $14\frac{1}{2}$ inches. Cylinder enclosed in opaque tube.
Copper,	K_4Cy_6Fe	1 part of metal detected in 4,000,000 of water.	1 part of metal detected in 11,750,000 of water.
"	NH_4HO	1,000,000 "	1,950,000 "
"	H_2S	4,150,000 "	15,660,000 "
Zinc,	NH_4HS	2,500,000 "	. . .
Arsenic,	H_2S	3,600,000 "	7,520,000 "
Lead,	K_2CrO_4	4,000,000 "	5,875,000 "
"	H_2S	100,000,000 "	196,000,000 "

BIOLOGIC EXAMINATIONS.

In a comprehensive sense the living organisms of water include representatives of all the great groups of animals and plants. The presence of any of the higher orders of organic forms may be taken as an indication of moderate purity, as these are absent from very foul water. From an analytic point of view, observation is limited to the determinations of those forms which are inappreciable to the unassisted eye. As far as regards some of the moderately complex organisms, such as the minute crustaceans, algæ, desmids, and even the amebæ, it may be said that while some general inferences as to the character and history of the water may be deduced from an identification of the specific forms, no definite sanitary signification can be attached to them. The ova of the entozoa might in some cases be detected by careful search, and would indicate recent pollution of a highly dangerous character.

Cohn (*Beitr. z. Biol. d. Pflanz.*) regards chlorophyl-producing plants (diatoms and green algæ), together with the infusoria that feed upon them, and species of entomostraca (*Daphnia* and *Cyclops*), when present in only moderate amounts, as indicating water not very rich in dissolved organic matter. Species of *Cladothrix*, *Crenothrix* and *Beggiatoa*, which are among the larger bacteria, and frequently appear as branching forms, indicate suspended organic matter; while dissolved organic matter in a state of active decomposition is indicated by the presence of ordinary bacteria (*Bacilli*, *Spirilla*), etc.

Cladothrix dichotoma withdraws iron from water, and fixes it, causing obstructions in iron water pipes. *Crenothrix Kühniana* Rabenhorst, is seen in water containing iron and sometimes causes a disagreeable odor.

The number of the higher forms present in any sample will depend very much upon the point at which it is collected, they being more numerous in the neighborhood of large plants and at the bottom and sides of streams. Under our present knowledge, no pathogenic power can be assigned to the higher forms of organic life, except entozoa, but their bodies after death may indirectly contribute to the rapid increase of the bacilli proper, by serving as food.

Several observers, notably Sedgwick and Rafter, have paid considerable attention to the recognition of the animal and vegetable forms in surface-waters. Some of these forms, while not apparently directly disease-producing, cause disagreeable odors and colors in water; in the warm season of the year when these waters are stored in reservoirs, considerable annoyance is felt by the users, and the engineer-in-charge is subjected to much criticism. For some years the American Water-works Association has had a special committee engaged in collecting data bearing upon this point, and devising methods for preventing the conditions. No satisfactory explanation or remedy has yet been offered by this committee, but it has been found that even crude filtration-methods, such as allowing the water to pass through a dike of porous soil before storing it in a reservoir, will diminish the tendency to these conditions. Cleansing a reservoir, disinfecting the inner surface, for instance, by whitewashing, has also improved the condition. (See Addenda, p. 136.)

A summary of Sedgwick's method, with some modifications by Dr. Williston, is given by the latter in a Report on Connecticut water-supplies for 1889-90.

In ordinary cases about 100 c.c. are employed. Sometimes it will be advantageous to use double this quantity, at other times much less. In rare cases the examination can be made upon unfiltered water. Originally sand was employed for a filter material, but Williston finds that precipitated silica made by decomposing silicon fluorid with water is more satisfactory. This precipitated silica is a commercial article, and its method of preparation is given in all the larger manuals of chemistry.

A small glass funnel with an even-calibered stem is selected, and the lower end of the stem plugged with a little absorbent cotton, upon which a layer three or four mm. deep of the filter-material is placed. The requisite volume of water is then allowed to filter through. The pledget of cotton is removed, and the filter-material is washed down with filtered or distilled water into a cell intended for microscopic examination. This cell is a glass plate accurately ruled, to which is attached a brass cell 50 mm. long by 10 mm. wide, of depth sufficient to hold about two c.c. of water. After the material has been allowed to distribute itself and settle in the cell, it is examined with a moderate power, and the different organisms in a varying number of the squares counted. Each organism may be counted by itself, if occurring in large numbers, the average of a few squares being sufficient for the purpose. Organisms less numerously represented may be counted by averaging a larger number of squares.

Filtering in this manner cannot be relied upon in all cases. Indeed, in most cases the unfiltered water also should be examined. Some of the minute unicellular organisms pass readily through the small extent of sand or precipitated silica, or even through filter paper.

It is not unlikely that the high-speed centrifugal apparatus now used in laboratories, associated with the employment of some fine precipitant will aid in these investigations. (See Addenda, p. 136.)

The introduction of the ova of entozoa into the human system by means of water is doubtless a very common occurrence, and cases have been reported. One of the most striking of these was the anemia occurring among the workmen in the St. Gothard tunnel, which was found to be due to the ingestion of the ova of a parasitic animal, *Anchylostomum duodenale.*

Culture Media.—For bacteriologic examinations, culture media prepared with gelatin or agar-agar, are generally used. In special cases potatoes and blood-serum are employed. In all cases solutions and vessels must be thoroughly sterilized. This is most easily accomplished in the Arnold steam sterilizer. It consists of a copper boiler, in the form of an inverted funnel, which communicates with the sterilizing chamber. A double casing is so arranged that the condensed steam falls into the pan and returns to the boiler. This pan should be about half-filled with water before starting. This apparatus is suitable not only for all sterilizations, but also for preparing solutions and for hot filtrations.

Meat-extract-peptone-gelatin:—

Water,	1000 c.c.
Meat extract,	5 grams
Gelatin,	150 "
Peptone,	10 "
Glucose,	2 "

The materials, which should be of good quality, are dissolved by heat and the solution rendered slightly alkaline by the addition of sodium carbonate or trisodium phosphate, added by small portions, stirring between each addition, and testing the liquid by placing a drop of it on red litmus paper. When the alkaline reaction appears, the liquid is filtered through good filter paper, the funnel and beaker being placed in the sterilizer, and a gentle steaming maintained in order to keep the gelatin liquid.

In place of meat extract, a solution may be made as follows: macerate 500 grams of finely minced lean meat in 1000 c.c. of water, for twenty-four hours, at a temperature not over 45° F. Strain the liquid, add five grams of common salt and the peptone, gelatin and glucose, and render feebly alkaline, as before. (See Addenda, p. 137.)

Agar-agar Mixture.—When cultures are to be conducted at a temperature above 68° F., agar-agar must be used as gelatinizing ingredient. The solution is prepared as given above, except that only 15 grams of agar-agar must be used and a much longer time will be required for its solution. The operation can be facilitated by soaking the agar-agar for twelve hours in a strong solution of common salt, or by the following treatment, which was first described by Macé:—

Twenty grams of agar-agar, cut fine, are steeped for twenty-four hours in 500 c.c. of water, containing six per cent. of hydrochloric acid, with occasional stirring. It is then well washed, and placed for a similar period in the same amount of water, to which six per cent. of ammonium hydroxid has been added. It is again washed thoroughly and added to 1000 c.c. of boiling water. It dissolves at once. The peptone, gelatin and salt can then be added, the mixture rendered feebly alkaline and filtered in the sterilizer.

Agar-agar Gelatin.—To secure the advantages of both agar-agar and gelatin, a solution may be made by dissolving 50 grams of gelatin and 7.5 grams of agar-agar in a liter of water, adding the peptone, etc., in the usual proportions, rendering alkaline, filtering and sterilizing. The proportions of agar-agar and gelatin must be adhered to closely or the gelatinization may be lumpy and incomplete.

Potato Culture.—Cultivation on potatoes was once much used as a method of distinguishing certain microbes. Large, sound potatoes should be selected, thoroughly washed, and cut into disks about five cm. in diameter and one cm. thick. These are placed in glass boxes (pomade boxes) which have lids with ground joint, and heated for about one-half hour in the sterilizer. Another method is to cut out cylinders with the aid of an apple-corer, or largest size cork-borer, slice these obliquely, and place them in test-tubes, which are then closed with cotton plugs, and sterilized. The latter method does not give a large surface, but the growth of any inoculation may be easily watched.

Collection of Samples.—Bacteriologic examinations are of little value unless promptly made on samples that have been collected with precautions against contamination. The inoculation of the culture-medium is best done at the source. If this is not possible, glass-stoppered bottles holding about 200 c. c., which have been thoroughly sterilized, with stoppers in place, in a hot air oven at 200° F., must be used for collection. They should be rinsed on the *outside* with water, dipped below the surface, the stopper withdrawn, and again inserted when the bottle is full. If these are to be transported any distance they should be packed in ice. For delivering the measured volume of water, a pipette sterilized in the hot-air oven should be used.

Culture Manipulations.—For the estimation and isolation of microbes in water, several methods may be employed, either the original plate-culture of Koch, Esmarch's "roll-culture," or a modification that Dr. Beam and I have used, which may be designated "bottle culture."

Plate-culture.—Test-tubes containing about 10 c. c. of nutrient jelly are plugged with cotton and steamed for fifteen minutes in the sterilizer on two successive days. In filling the tubes, care must be taken that none of the jelly touches the upper part, where it can come in contact with the cotton plug. After sterilization, the tubes should be put aside for a few days, to determine if all spores are destroyed. If no development of microbes takes place in this time, the jelly is ready for use. It should be melted at a low temperature in a water-bath, the cotton plug removed with a twisting motion and a measured volume of water introduced by means of a sterilized pipette. The plug is immediately replaced and the liquids mixed by shaking, taking care not to soil the cotton. The quantity of water to be added will depend on the number of microbes supposed to be present. If the amount is probably quite small, one c. c. should be taken; if it is probable that a large number is present, one c. c. of the sample should be diluted to 10 c. c. with sterilized water, and one c. c. of this mixture used. In most cases it will be well to make several cultures, using different proportions of water. After the jelly and water have been mixed the liquid is poured out on a glass plate and allowed to set. The principal difficulty in the method arises from the liability to contamination from the air during this part of the process. The glass plates should, of course, be thoroughly sterilized. They are usually set upon glass benches, and placed in a so-called moist chamber, which consists of two cylindrical glass dishes, one fitting within the other. The benches and plates are placed in the inner dish. To hasten the setting of the jelly, it is necessary that the plates should be cold, and it will be best to have them resting in a level position on a flat tin bottle filled with ice water, or, if ice is not at hand, a recently prepared mixture of one part of ammonium nitrate with two parts of water. The coated plates must be protected from dust while being cooled, and as soon as possible must be transferred to the moist chamber, which, as well as the benches, should have been previously thoroughly washed with recently boiled water. The bottom should be covered with a sheet of moist filter-paper. The plates are kept in the chamber for several days. Each living microbe that is capable of growing in the medium will become the center of a colony of its own kind, which will soon become visible to the naked eye. The number of colonies may be counted as soon as they are distinct. When the number is very large, an approximate counting may be made by means of a glass plate with lines ruled so as to divide it into a considerable number of equal squares. This is placed over the culture, and the number of colonies in several of the squares counted, the result averaged and multiplied by the number of squares that the entire culture covers.

Roll Culture.—Instead of pouring the jelly on a plate, it may be spread in a uniform layer on the inner wall of the test-tube, taking care not to allow it to touch the cotton. The jelly may be rapidly hardened by rolling the tube in contact with ice. The method avoids all danger of contamination with dust, and is very simple in manipulation, but is open to the objection that when bacilli are present which liquefy the gelatin, as will generally be the case, the fluid will run down and inoculate other portions of the layer.

Bottle Culture.—To avoid the disadvantage of the above methods, it will be found convenient to employ flat rectangular bottles of the form known technically as the "Blake" Bottle. Those of 8 or 10 ounce capacity are the best, and it is well to select such as have as uniform a thickness of glass as possible. They should be thoroughly cleaned and sufficient of the jelly put in to form a thin layer on one of the broader sides when the bottle is placed horizontally. The mouth is then closed by a cotton-plug and the bottle sterilized as described in connection with plate culture. To make a culture, the jelly is melted at a low temperature, the measured volume of water added, the cotton plug replaced, and after mixing, the bottle is placed in a horizontal position for the usual time.

General Character of the Microbes in Natural Waters.—The microörganisms of natural waters are principally included in the genera *Bacillus* and *Spirillum*, especially the former. Micrococci and moulds are rare, and when they appear on the culture-plate, are generally due to contamination by dust.

Microbes are distinguished according to the conditions favorable to their growth, as follows:—

Saprophytic. Growing on dead matter.
Parasitic. Growing only on living matter.
Aërobic. Requiring free oxygen.
Anaërobic. Not requiring free oxygen.

When the organism possesses the power of adapting itself to different conditions, the term *facultative* is applied; when it can grow only under special conditions, the term *obligatory* is applied.

Microbes are also differentiated by the effect which they produce upon the culture-medium. Some species rapidly or slowly liquefy the jelly with evolution of foul smelling gases; others—chromogenic microbes—produce characteristic colors. Many do not produce any positive modification, and for purposes of distinction it is usual to transfer portions of the colonies to other culture-media. Such special cultures are obtained by taking up a portion of the colony on the end of a wire which has been just sterilized by heating to redness and inoculating the prepared medium.

Cultivation in Absence of Oxygen.—Several of the most frequently occurring microbes are obligatory anaërobes, and will grow, therefore, only in an atmosphere deprived of free oxygen. Carbon dioxid has been found to act unfavorably upon their development. The most suitable atmosphere is one of pure hydrogen. Cultivation in such an atmosphere may be secured by constructing the moist chamber so as to permit its being filled with hydrogen and sealed, or by the use of Liborius' tube, Fig. 8.

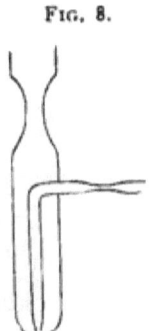

Fig. 8.

The tube is charged with nutrient jelly, plugged with cotton, sterilized, inoculated with the material to be tested, the jelly maintained in a liquid condition by a very gentle heat and a current of pure hydrogen passed through the side tube until all air is expelled. The test-tube and side-tube are then sealed quickly at the narrow portions, in the blow-pipe flame, and the jelly allowed to solidify.

Staining.—The differentiation of the various species of microbes may also be accomplished by staining with various anilin colors. These methods are, however, more generally applicable to pathologic work, that is, to the staining of microbes in tissues.

Indol Reaction.—Indol, C_8H_7N, more properly indin, is a weak base, which is a product of the growth of many species of microbes, and the detection of it may, therefore, be utilized as a differentiation test. S. Kitasato (*Zeit. f. Hyg.*, VII, 518) gives the following method for performing the test:—

Ten c. c. of an alkaline-peptone-meat infusion (without

gelatin), which has been previously inoculated with the microbes to be tested, and kept for twenty-four hours at blood heat, are treated with one c. c. of solution of pure potassium nitrite (0.02 grm. in 100 c. c.) and then with a few drops of concentrated sulphuric acid. In the presence of indol a rose or deep-red color is developed. *Spirillum choleræ* and *Bacillus coli-communis* give the reaction strongly; *S. Finkleri* feebly; the so-called *B. typhosus* does not give it.

An effort has been made to distinguish dangerous microbes by employing crucial conditions of cultivation which are believed to eliminate the harmless forms. One of the most complete methods is that used in the Hygienic Laboratory of the University of Michigan, under the supervision of Dr. Vaughan:—

The water is collected in sterilized bottles, and beef-tea tubes are inoculated with it, from one drop to one cubic centimeter being added to each tube. The inoculated tubes are kept at 38° C. for twenty-four hours.

Many germs found in drinking-water will not grow at so high a temperature. It is assumed that a germ which will not grow at the temperature of the human body cannot possibly induce disease. Of the germs which do grow at 38°, or higher, some are toxicogenic to animals, while others are not. Whether the germs in a given sample are toxicogenic or not, is determined by injecting the prepared cultures subcutaneously or intra-abdominally in rats, guinea-pigs, rabbits, or mice.

Waters not containing toxicogenic germs growing at 38° or higher are reported as safe.

Waters containing toxicogenic germs growing at 38° C. are condemned. In testing virulence, a germ is not pronounced toxicogenic unless it be found growing and multiplying in the animal inoculated with it.

The Franklands ("Microörganisms in Water"), after reviewing at length the various methods that have been suggested for differentiating the typhoid-bacillus, show that most of them are quite insufficient. A routine treatment is, however, suggested as follows:—

The water is freed from the ordinary water-bacteria by preliminary culture in Parietti's solution. This is a phenol-broth prepared by mixing 10 c. c. of neutral bouillon with 0.25 c. c. of the following solution:—

Phenol,	5 grams.
Hydrochloric acid,	4 "
Water,	100 "

This solution is sterilized by heating it for twenty-four hours at 37° C. The tubes are then inoculated with from one to ten drops of the water-sample, which must be thoroughly mixed with the broth, and again kept in the sterilizer for not less than forty-eight hours. If any of the tubes appear turbid after the treatment, they should be submitted to plate-cultivation, and any colonies which resemble those produced by the typhoid-bacillus should be further studied by inoculation into gelatine tubes to observe the gas-producing test, into milk to note if coagulation occurs, and by cultivation in broth to determine if indol-reaction will be obtained.

Hazen and White ("Manual of Bacteriology," Sternberg, p. 354) cultivate the water in agar-agar tubes at 40° C. for several days, which prevents the growth of many of the common water-microbes.

Vaughan ("Proceedings American Association of Physicians, 1892") investigated the application of these principles to the isolation of a specific bacillus, but did not obtain satisfactory results.

From the Franklands' work I take the description of methods used for detection of the *Spirillum choleræ* in water. According to Koch, 100 c. c. of water is mixed with one per cent. each of peptone and common salt and the mixture kept at 37° C. At intervals of ten, fifteen, and twenty hours plate-cultivations are made of portions of the sample. Any colonies which appear on the plates should be reinoculated into fresh culture-media and be tested after a time for indol.

Schottelius recommends the following: The water is mixed with twice its volume of alkaline sterile bouillon, and kept at a temperature of 30–40° C. for twelve hours. An extensive multiplication of the spirillum occurs on the surface of the liquid, and the examination of the pellicle will show many of them.

Klein (*Br. Med. Jour.*, Oct. 14, 1894) has pointed out that in most of the methods for isolating pathogenic microbes, the volume of water used is too small. He suggests the employment of at least 1000 c. c. Obviously this volume will give a better sampling than the few drops or 1 c. c. generally used.

TECHNIC EXAMINATIONS.

GENERAL QUANTITATIVE ANALYSIS.

Silica, Iron, Aluminum, Manganese, Calcium, and Magnesium.—One liter of the water acidified with hydrochloric acid is evaporated to complete dryness, best in a platinum dish, the residue treated with hydrochloric acid and water, and the separated *silica* filtered, washed, dried, ignited in a platinum crucible, and weighed.

To the filtrate, previously boiled with a few drops of strong nitric acid, slight excess of ammonium hydroxid is added, the liquid boiled several minutes, the precipitate collected, washed thoroughly with boiling water, dried, ignited, and weighed. It consists of Fe_2O_3 and Al_2O_3. It also contains all the phosphates and some manganese if much is present in the water. In such cases the precipitate before drying is re-dissolved in hydrochloric acid and neutralized with a dilute solution of ammonium carbonate until the water almost becomes turbid. It is then boiled and the precipitate, now free from manganese, washed, dried, ignited, and weighed. The iron may be determined by dissolving the precipitate in strong hydrochloric acid and employing the colorimetric method described on page 38.

If no manganese or only traces are present, the filtrate from the iron is mixed with sufficient ammonium chlorid to prevent the precipitation of the magnesium, ammonium hydroxid, and then ammonium oxalate added in quantity sufficient to precipitate the calcium and to convert all the magnesium into oxalate, and thus hold it in solution. The precipitate contains all the calcium and some of the magnesium. If the magnesium is present only in relatively small quantity the amount carried down may be disregarded; otherwise a second precipitation should be made as follows: The solution is allowed to stand until the precipitate has subsided; this will require some hours. The supernatant liquid is poured off through a filter, the precipitate washed by decantation, then dissolved in hydrochloric acid, water added, then ammonium hydroxid and a small quantity of ammonium oxalate. After the calcium oxalate has thoroughly subsided it is filtered off, washed, and dried. If quite small in amount it is placed with the filter in a weighed platinum crucible, ignited over the Bunsen burner for a short time, and then over the blast lamp for from five to fifteen minutes. The calcium is thus obtained in the form of oxid, which is allowed to cool in the desiccator and weighed. The weight thus obtained multiplied by 0.714 gives the weight of *calcium*. When the amount of precipitate is large, it is better to remove it from the filter, and heat it just short of redness until it assumes a grayish tint. It then consists of calcium carbonate. To this is added the ash of the filter. The weight of the calcium carbonate multiplied by 0.4 gives the weight of calcium.

The filtrates are mixed, slightly acidified with hydrochloric acid, concentrated and cooled, ammonium hydroxid and sodium phosphate added in excess, stirred briskly and allowed to stand in the cold for about twelve hours. The precipitated ammonium magnesium phosphate is brought upon a filter, that adhering to the sides of the vessel being dislodged by rubbing with a glass rod tipped with a piece of clean rubber tubing. It is washed with a solution made by mixing one part of the ammonium hydroxid of 0.96 sp. gr. with three parts of water. The precipitate is dried, transferred to a platinum crucible, the filter ashed separately and added to it, and the whole heated at first gently and then to intense redness for several minutes. After cooling, it is weighed. It consists of magnesium pyrophosphate; the weight multiplied by 0.218 gives the weight of *magnesium*.

Manganese, if present in appreciable quantity, is separated before the precipitation of the calcium, as follows:

The filtrate from the iron precipitate is slightly acidulated with hydrochloric acid, concentrated, and the manganese precipitated as sulphid by colorless or slightly yellow solution of ammonium sulphid. The flask, which should be nearly full, is stoppered, allowed to rest in a moderately warm place until the precipitate has thoroughly settled, filtered, washed with dilute ammonium sulphid water and purified by dissolving in a small quantity of hydrochloric acid and reprecipitating with ammonium sulphid. It is filtered off, washed as before, dried, placed in a weighed porcelain crucible, covered with a little sulphur and ignited in a current of hydrogen introduced into the crucible by a tube passing through a hole in the cover. The pure manganese sulphid thus obtained is allowed to cool and weighed. The weight multiplied by .632 gives *manganese*.

Sulphates.—500 c. c. of the clear water are slightly acidulated with hydrochloric acid, heated to boiling, and barium chlorid solution added in moderate success. The precipitate is allowed to subside completely, collected upon a filter, washed thoroughly, dried and incinerated. It is $BaSO_4$; the weight multiplied by 0.412 gives SO_4. If the proportion of SO_4 is very low, it will be advisable to concentrate the water to one-fifth or one-tenth its bulk before precipitating.

Control. Potassium, Sodium, and Lithium.— From 250 to 1000 c. c. of the water, according to the amount of solid matters present, are evaporated to dryness in a platinum dish, and the residue treated with a small amount of water and sufficient dilute sulphuric acid to decompose the salts present. The dish should then be covered and placed upon the water-bath for five or ten minutes, after which any liquid spurted on the cover is washed into the dish, the whole evaporated to dryness and heated to redness. A few drops of ammonium carbonate solution should then be mixed with the residue, and the ignition repeated to insure the removal of the last portions of free acid. In the majority of cases the only basic elements present in considerable quantity are calcium, magnesium, and sodium. The *sodium* may be determined indirectly, therefore, by calculating from the amount of Ca and Mg found, the calcium and magnesium sulphate in the residue, and subtracting this sum, together with the silica, from the total residue.

For the determination of potassium and sodium in ordinary well and river waters, not less than two liters should be employed. When lithium is to be determined, it is generally necessary to use much more. In any case, as the alkalies are to be weighed as chlorids, it is advisable, if notable amounts of sulphates are present, to precipitate them by addition of barium chlorid.

The water is evaporated to about 200 c. c., a slight excess of calcium hydroxid added to the hot liquid—generally 3 c. c. of thin milk of lime will be sufficient—and the heat continued for several minutes. It is then washed into a 250 c. c. flask, disregarding the insoluble portion adhering to the dish, which, however, should be thoroughly washed, and the washings added to the flask. After cooling, the flask is filled up to the mark with distilled water, thoroughly mixed, the precipitate allowed to settle, and the liquid filtered through a dry filter. 200 c. c. of the filtrate are measured into another 250 c. c. flask, ammonium carbonate and ammonium oxalate added, filled with water up to the mark, mixed, allowed to settle, filtered through a dry filter, 200 c. c. of the filtrate measured off and evaporated to *thorough dryness* in a platinum crucible, heating very cautiously at the last stages to avoid loss by spurting. The low-temperature burner is suited for this purpose. The crucible is now covered and cautiously heated to dull redness, cooled and weighed. The residue contains the potassium, lithium and sodium as chlorids. It contains, sometimes, also, traces of magnesium, which may be removed by treating again with lime and ammonium carbonate and oxalate. It is frequently of advantage, in evaporating these saline solutions, to add, when the solution becomes concentrated, several c. c. of strong hydrochloric acid. This precipitates the greater portion of the salts in a finely granular condition, and renders loss by spurting less liable to occur.

If potassium and sodium chlorids only are present, the residue is dissolved in a small quantity of water, an excess of a concentrated neutral solution of platinum chlorid added, evaporated to small bulk at a low heat on the water bath, some 80 per cent. alcohol added, allowed to stand, the clear liquid decanted off on a small filter and the residue washed in this way several times by fresh small portions of 80 per cent. alcohol. The precipitate is then washed on to the filter with alcohol, washed again with 80 per cent. alcohol, thoroughly dried and transferred as far as possible to a watch glass. The small portion on the filter is dissolved off and the solution placed in a weighed platinum dish and evaporated to dryness. The main portion on the watch glass is then added, and the whole dried to a constant weight at about 260° F., cooled and weighed. The weight thus found multiplied by .30 gives the weight of *potassium chlorid*. This subtracted from the combined weight of the chlorids gives the weight of *sodium chlorid*.

Lithium, if present, is best separated before the treatment with platinum chlorid. The following method, devised by Gooch, gives good results: To the concentrated solution of the weighed chlorids, amyl alcohol is added and heat applied, gently at first, to avoid bumping, until the water disappears from the solution and the point of ebullition becomes constant at a temperature which is approximately that at which the alcohol boils ($270°$ F.), the potassium and sodium chlorids are deposited and the lithium chlorid is dehydrated and taken into solution. The liquid is then cooled and a drop or two of strong hydrochloric acid added to reconvert traces of lithium hydroxid in the deposit, and the boiling continued until the alcohol is again free from water. If the amount of lithium chlorid be small, it will be found in the solution and the potassium chlorid and sodium chlorid in the residue, excepting traces which can be allowed for. If the lithium chlorid exceed ten or twenty milligrams the liquid may be decanted, the residue washed with amyl alcohol, dissolved in a few drops of water and treated as before. For washing, amyl alcohol, previously dehydrated by boiling, is to be used and the filtrates are to be measured apart from the washings. In filtering, the Gooch filter with asbestos felt may be used with advantage, applying gentle pressure by the aid of the filter pump. The crucible and residue are ready for weighing after gentle heating over the low-temperature burner. The weight of the insoluble chlorids is to be corrected by adding .00041 for every 10 c. c. of amyl alcohol in the filtrate, exclusive of the washings, if only sodium chlorid be present; .00051 for every 10 c. c. if only potassium chlorid, and .00092 in the presence of both these chlorids.

The filtrate and washings are evaporated to dryness in a platinum crucible heated with sulphuric acid, the excess driven off, and the residue ignited to fusion, cooled and weighed. From the weight is to be subtracted, for each 10 c. c. of filtrate, .0005, .00059, or .00109, according as only sodium chlorid, potassium chlorid, or both were present in the original mixture.

Hydrogen Sulphid.—The following method is taken from Sutton's "Volumetric Analysis:"—

Reagents Required:—

Centinormal Iodin.—Dry, commercial iodin is intimately mixed with one-fourth its weight of pure potassium iodid and gently heated between two clock-glasses by resting the lower on a hot plate. The iodin sublimes in a perfectly pure condition. It is allowed to cool under the desiccator, 1.265 grams weighed out, together with 1.8 grams of pure potassium iodid, dissolved in about 50 c. c. of water and the solution made up exactly to a liter. The liquid must not be heated, and care should be taken that no iodin vapor is lost. One c. c. is equivalent to .00017 H_2S. The solution is best prepared in stoppered bottles, which should be completely filled and kept in the dark. It will not even then keep very long, and should be standardized by titration with a weighed amount of pure sodium thiosulphate, which should be powdered previous to weighing, and pressed between filter paper to absorb any moisture. 50 c. c. of the iodin solution, when of full strength, will require 0.124 gram of sodium thiosulphate.

Starch Indicator.—See page 48.

Analytic Process:—

Ten c. c., or any other necessary volume of the iodin solution, is measured into a 500 c. c. flask, and the water to be examined added until the color disappears. Five c. c. of starch liquor are then added and the iodin solution run in until the blue copper appears; the flask is then filled to the mark with distilled water. The respective volumes of iodin and starch solution, together with the added water, deducted from the 500 c. c. will show the volume of water actually titrated by iodin. A correction should be made as follows for the excess of iodin required to produce the blue color: Five c. c. starch solution are made up with distilled water to 500 c. c., iodin run in until the color matches that in the test, and the volume of iodin solution so used subtracted from the figure obtained in the first titration.

Hardness. CO_3 in Normal Carbonates.—

Waters containing considerable quantities of calcium and magnesium are said to be hard. Since the solution of calcium and magnesium carbonate in water depends partly upon the presence of carbon dioxid, boiling precipitates the greater portion of the carbonates, the result being to diminish the hardness, *i. e.*, to soften the water. Magnesium and calcium sulphates and chlorids are not precipitated in this way. Hardness, therefore, is divided into two classes, temporary and permanent, the former being that which may be removed by boiling. The process here described is due to Hehner.

Reagents Required:—

Standard Sodium Carbonate.—1.06 grams of recently ignited pure sodium carbonate are dissolved in water and the solution diluted to 1000 c. c. One c. c. = .00106 gram Na_2CO_3, equivalent to .001 gram $CaCO_3$.

Standard Sulphuric Acid.—One c. c. of pure concentrated sulphuric acid is added to about 1000 c. c. of water. 50 c. c. of the standard sodium carbonate are placed in a porcelain dish, heated to boiling, a few drops of a solution of phenacetolin or lacmoid added, and the sulphuric acid cautiously run in from a burette until the proper change of color occurs. From the figure thus obtained, the extent to which the acid should be diluted in order to make one c. c. of the sodium carbonate equivalent to one c. c. of the acid may be calculated. The proper amount of water is then added and the solution verified by again titrating.

Analytic Process :—

Temporary Hardness.—100 c. c. to 250 c. c. of the water tinted with the indicator are heated to boiling, and the sulphuric acid cautiously run in until the color change occurs. Each c. c. required will represent one part of calcium carbonate or its equivalent per 100,000 parts of the water.

Permanent Hardness.—To 100 c. c. of the water is added an amount of the sodium carbonate solution more than sufficient to decompose the calcium and magnesium sulphates, chlorids and nitrates present; usually a bulk equal to the water taken will be more than sufficient. The mixture is evaporated to dryness in a nickel or platinum dish, and the residue extracted with distilled water. The solution is filtered through a very small filter, and the filtrate and washings titrated hot with sulphuric acid as above; or 25 c. c. of distilled water may be poured on the residue, and the solution obtained filtered through a dry filter, the filtrate measured and titrated. The difference between the number of c. c. of sodium carbonate used and the acid required for the residue will give the permanent hardness.

If the water contains sodium or potassium carbonate there will be no permanent hardness, and there will be more acid required for the filtrate than the equivalent of the sodium carbonate added. From this excess the quantity of sodium carbonate in the water may be determined.

Since any alkali-carbonate in the water would be erroneously calculated as temporary hardness by the direct titration, the equivalent, in terms of calcium carbonate, of the alkali carbonate present should be deducted from the figure given by the titration in order to get the true temporary hardness.

The total CO_3 in normal carbonates is given by the direct titration of the water with dilute sulphuric acid. One c. c. of the acid is equivalent to .0006 gram of CO_3.

Free Carbonic Acid.—The following process, due to Pettenkofer, is taken from Sutton's "Volumetric Analysis:"—

A stoppered bottle of known capacity, about 150 c. c., is filled at the source by submergence, or, if taken from a faucet, by allowing the stream to run in with full force for some minutes, the nozzle being inserted into the neck of the bottle. 50 c. c. of the water are then quickly removed by a pipette, and the following solutions immediately added: Three c. c. of a saturated solution of calcium chlorid and two c. c. of a saturated solution of ammonium chlorid; 45 c. c. of clear calcium hydroxid solution of known strength are added, the flask well corked, the liquids mixed, and set aside for at least twelve hours, to allow the calcium carbonate formed to settle and become crystalline and insoluble. An aliquot part (50 to 100 c. c.) of the clear liquid is then drawn off and titrated with decinormal acid, using phenacetolin or lacmoid as indicator, and from the amount required the entire proportion of calcium hydroxid unacted upon can be determined. This being deducted from the amount originally added, and the remainder multiplied by .0022, will give the weight of carbonic acid in the water in excess of that existing as normal carbonates.

Boric Acid.—*Detection.*—Add to one liter of the water sufficient sodium carbonate to render it distinctly alkaline. Evaporate to dryness, acidify with hydrochloric acid, moisten a slip of turmeric paper with the liquid, and dry it at a moderate heat. In the presence of boric acid the paper will assume a distinct brown-red tint.

Quantitive determinations of boric acid are rarely required. It exists in many of the geyser-waters of Yellowstone National Park, and doubtless in similar waters in other parts of the world. A careful investigation into the methods of quantitive determination was published by Hehner (*The Analyst*, August, 1891) and a modification of Gooch's method is described by Moissan (*Jour. Soc. Chem. Ind.*, April 1, 1895).

SPECTROSCOPIC EXAMINATION.

For the ordinary spectroscopic examination of a water a simple apparatus will suffice. The arrangement figured in the cut (Fig. 9) is a small direct-vision spectroscope, held in a universal stand, with an adjustable burner as the source of heat. The entire apparatus does not cost over $15.00, and will be found convenient and efficient.

For the examination, a liter or more should be evaporated nearly to dryness, a little hydrochloric acid being added near the end of the process, the residue placed in a narrow strip of platinum foil having the sides bent so as to retain the liquid, and heated in the flame. While this method will be sufficient in many cases, a far better plan is to separate the substance sought for in a state of approximate purity and then examine with the spectroscope. Very small traces of lithium, for instance, may be detected as follows: To about a liter of the water sufficient sodium carbonate is added to precipitate all the calcium and magnesium, and the liquid boiled down to about one-tenth its bulk; it is then filtered, the filtrate rendered slightly acid with hydrochloric acid and evaporated to dryness. The residue is boiled with a little alcohol, which will dissolve out the lithium chlorid. The alcoholic solution is evaporated to dryness, the residue taken up with a little water and tested in the flame.

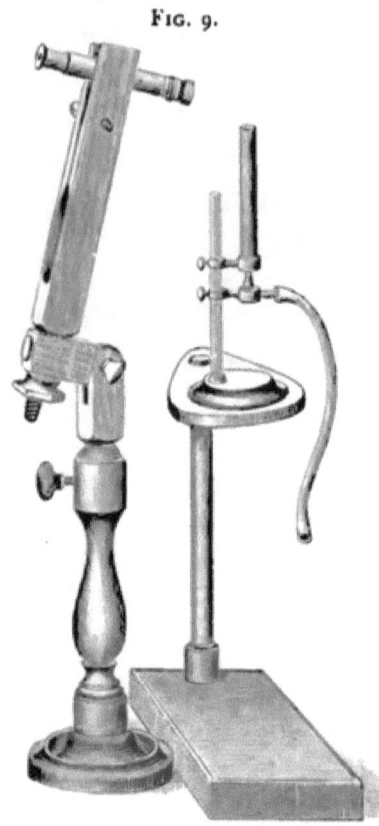

FIG. 9.

In order to identify with certainty any line which may be obtained, it is only necessary to hold in the flame at the same time a wire which has been dipped in a solution of the substance supposed to be present, and to note whether the lines produced by it and the material under examination are identical.

SPECIFIC GRAVITY.

In the great majority of cases the determination of specific gravity is not essential. Ordinary river, spring and well waters contain such small proportions of solid matter that it is usually the practice to take a measured volume and to assume its weight to be that of an equal bulk of pure water. If the proportion of solids be high, a determination of the specific gravity may be desirable. For this purpose the specific gravity bottle may be used. This consists merely of a small flask provided with a finely perforated glass stopper. The bottle is weighed first alone, then filled with distilled water at 60° F., and finally with the water under examination at the same temperature. In filling the bottle, the liquid is first brought to the proper temperature, the bottle completely filled, the stopper inserted, and the excess of water forced out through the perforation and around the sides of the stopper, carefully removed by bibulous paper. The weight of the water examined divided by the weight of the equal bulk of distilled water at the same temperature gives the specific gravity.

Another method, and one which gives very satisfactory results, is by the use of a plummet. This may conveniently consist of a piece of a thick glass rod of about 10 c. c. in bulk, or of a test-tube weighted with mercury and the open end sealed in the flame. The plummet is suspended to the hook of the balance by means of a fine platinum wire and its weight ascertained. It is then immersed in distilled water at 60° F., and the loss in weight noted. The figure so obtained is the weight of a bulk of water equal to that of the plummet. This having been determined, the specific gravity of any water may be found by immersing in it the plummet and noting the loss in weight. This, divided by the loss suffered in pure water, gives the specific gravity.

INTERPRETATION OF RESULTS.

STATEMENT OF ANALYSIS.

The composition of water is generally expressed in terms of a unit of weight in a definite volume of liquid, but much difference exists as to the standard used. The decimal system is very largely employed, the proportions being expressed in milligrams per liter, nominally parts per million; or in centigrams per liter, nominally parts per hundred thousand. Not infrequently the figures are given in grains per imperial gallon of 70,000 grains, or the U. S. gallon of 58,328 grains. Much more rarely grains per quart, parts per thousand, per cent., or other inconvenient ratios are employed. In this work the composition is always expressed in milligrams per liter. This ratio is practically equivalent to parts per million, except in case of water very rich in solids, a liter of which weighs notably more than one million milligrams. Factors for converting the different ratios are given at the end of the book.

From the analysis of a water it is rarely possible to ascertain the exact arrangement of the elements determined, but it is the custom to assume arrangements based upon the rule of associating in combination elements having the highest affinities, modifying this system by any inferences derived from the character or reactions of the water itself. It has been demonstrated that, even in the case of mixtures of salts producing no insoluble substances, partial interchange of the basylous and acidulous radicles takes place. In a solution of sodium chlorid and potassium sulphate sodium sulphate and potassium chlorid will be found, as well as the original salts. When the conditions are rendered more complex by the addition of other substances, it is obviously impossible to determine the exact arrangement. In view of these facts, it is preferable to express the composition of a water by the proportion of each element or radicle present. In this way a water containing K_2SO_4, will be expressed in terms of K and SO_4, respectively. In the case of bodies like CO_2 and SiO_2, which may possibly exist free in the water, their proportion is expressed as such. It frequently occurs that the characteristics of some of the compounds in a water are sufficiently marked to indicate their presence, and there can be no objection to suggesting, in connection with the analytic statement, the inferences which may thus be drawn.

The organic matters, or derived products, are best stated in terms of the nitrogen which they contain, thus permitting a comparison of the different stages of decomposition. It is inadvisable to represent the amount of unchanged organic matter in terms of oxalic acid, as has been suggested, or to express the nitrogen in terms of albumin, or any other supposititious compound.

SANITARY APPLICATIONS.

Judgment upon the analytic results from a given sample of water depends upon the class to which it belongs, and to the particular influences to which it has been subjected. A proportion of total solids which would be suspicious in a rain or river water, would be without significance in that from an artesian well. On the other hand, a subsoil water of unobjectionable character would contain a proportion of nitrates which would be inadmissible in the case of a river or deep water. Location has also much bearing in the case; subsoil waters near the sea will be found to contain, without invoking suspicion, proportions of chlorin which would be ample to condemn the same sample if derived from a point far inland. Hence the importance of recording, at the time of collection, all ascertainable information as to the surroundings and probable source of the water.

Analyses of surface-waters have little value, unless supplemented by a careful survey of the watershed to determine sources of pollution. Such survey will often discover conditions sufficient to condemn the supply, even though the analyses may be satisfactory. Indeed, it may be taken as a fundamental principle, that no water-supply derived from streams flowing through a populated district will be safe for use unless efficiently filtered.

Color, Odor, and Taste.—Water of the highest purity will be clear, colorless, odorless, and nearly tasteless. While in some cases a decided departure from this standard may give rise to suspicion, analytic observations are necessary to decide the point. Water highly charged with mineral matters will possess decided taste, vegetable matters may communicate distinct color; but, on the other hand, it may be highly contaminated with dangerous substances and give no indications to the senses. Well-waters occasionally become offensive in odor, from penetration of tree roots. The odor often recalls that of hydrogen sulphid. Sulphids are, indeed, often formed in such cases by the abstraction of oxygen from sulphates under the influence of microbes. Such waters are often used without apparent injury, but it is probable that if direct pollution occurs, the danger would be enhanced by the presence of the vegetable matter. Miquel has described a bacillus, under the name *B. sulphydrogenus*, which produces hydrogen sulphid readily. Some of the common putrefactive bacilli doubtless have this power also, largely through the influence of the hydrogen liberated by them. In waters containing hydrogen sulphid, species of beggiatoa, especially *B. alba*, thrive, and decomposing the sulphid, become impregnated with sulphur. Natural sulphur waters frequently contain these organisms, as do, also, waste-waters containing sulphid.

Surface waters collected in reservoirs or ponds often become very offensive from the growth of algæ, but apart from the disgust created by the water it is not known that any harmful results occur to those using it.

Turbidity may be due to several causes, of different degrees of danger, but is always objectionable.

Total Solids.—Excessive proportions of mineral solids, especially of marked physiological action, are known to render water non-potable, but no absolute maximum or minimum can be assigned as the limit of safety. Distilled water and waters very highly charged with mineral matter have been used for long periods without ill effects. The popular notion that the so-called hard waters conduce to the formation of urinary calculi is not borne out by surgical experience nor statistical inquiry. Many urinary calculi are composed of uric acid, and are the results of disorders of the general nutritive functions.

Sanitary authorities have fixed an arbitrary limit of total solids of about six hundred parts per million, but many artesian waters in constant use exceed this. An instance is found in the well on Black's Island, near Philadelphia, given in the table of analyses, which contains nearly twelve hundred parts per million, is very agreeable in taste, and has been in constant use for some years by a number of persons without injury. The assertion that water to be wholesome must contain an appreciable proportion of total solids is also not demonstrated by clinical experience. A discussion of the effects of special mineral ingredients, *e. g.*, magnesium sulphate, ferrous carbonate, etc., belongs to general therapeutics.

The odor produced on heating the water residue is often of much use in detecting contamination. Odors similar to those produced by heating glue, hair, rancid fats, urine, or other animal products, will give rise to grave suspicion. On the other hand, a more favorable judgment may be given when the odor recalls those given off in the heating of non-nitrogenous vegetable materials, such as wood-fibre.

Poisonous Metals.—The proportion of iron in water constantly used for drinking purposes should not much exceed three parts per million. Lead, copper, arsenum, and zinc must be considered dangerous in any amount, though it appears that zinc and copper, being least cumulative, are rather less objectionable in minute amount than the others. Concerning the limit of safety with manganese and chromium very little is known, but their presence in appreciable quantity must be looked upon with suspicion.

Chlorids and Phosphates.—Chlorids—principally sodium chlorid—and phosphates are abundantly distributed in rocks and soils, and find their way into natural waters; but while the former are freely soluble, and remain in undiminished amount under all conditions to which the water is subjected, all but minute amounts of the latter are either precipitated or removed by the action of living organisms. Surface and subsoil waters ordinarily contain but a few parts per million. Both chlorids and phosphates being constant and characteristic ingredients of animal excretions, it is obvious that an excess of them in natural waters, unless otherwise accounted for, will suggest direct contamination. Proximity to localities in which sodium chlorid is abundant, such as the sea or salt deposits, will deprive the figure for the chlorin of diagnostic value, nor can any indication of sewage or other dangerous pollution be inferred from high proportion of chlorin in deep waters. Further, it has been shown that the proportion of chlorin in uncontaminated waters is tolerably constant, while in water subjected to the infiltration of sewage the chlorin undergoes marked variation in amount. In most cases, therefore, a correct judgment can only be attained by comparison with the average character of the waters of the same type in the district, and by examination at intervals of the water in question.

As regards phosphates, Hehner, who has published a series of analyses, states that the presence of more than 0.6 parts per million—calculated as PO_4—should be regarded with suspicion. On the other hand, the absence of phosphates affords no positive proof of the freedom from pollution.

Nitrogen from Ammonium Compounds.—Ammonium compounds are usually the results of the putrefactive fermentation of nitrogenous organic matter; they may also be the product of the reduction of nitrites and nitrates in presence of excess of organic matter. In either case, therefore, they suggest contamination. Deep waters often contain an excess of ammonium compounds, derived, in large part, from the reduction of nitrates. Their presence here is hardly ground for adverse judgment, since the water, even though originally contaminated, has undergone extensive filtration and oxidation, its organic matter converted into bodies presumably harmless, and microbes have perished. Such waters, indeed, usually show only traces of unchanged organic matter.

Rain water often contains large proportions of ammonium compounds; but here, also, the fact cannot condemn the water, since it does not indicate contamination with dangerous organic matter.

The evolution of ammonia in the distillation of rain water often continues indefinitely, the larger portion passing over in the first distillates, but small quantities being present even after the distillation has been much prolonged. The same continuous evolution of ammonia is noted in waters containing urea, but in this case a larger proportion is collected in the earlier distillates, nearly all coming over before one-half the water has been distilled. Fox gives the following figures as ratios obtained in the analysis of two samples, one of rain water collected from a roof and therefore impure, and the other of a water containing urine:—

	Rain Water.		Urine Water.
1st distillate,	.35	.38	parts per million.
2d "	.25	.14	" "
3d "	.12	.065	" "
4th "	.09	.035	" "
5th "	.09	. .	" "
6th "	.04	. .	" "
7th "	.03	. .	" "
	.97	.620	

Nitrogen by Alkaline Permanganate (Nitrogen of "albuminoid ammonia").—A large yield of ammonia by boiling with alkaline potassium permanganate will, of course, point to an excess of nitrogenous organic matter. The inferences to be drawn depend upon the origin and condition of the organic material. If animal, the water may at once be condemned as unsafe. Waters containing excessive amounts even of vegetable matter are not free from objection, since they have frequently caused persistent diarrhea. If the organic matter, whether animal or vegetable, is in a state of active decomposition, it is doubly objectionable. Mallet has called attention to the fact that such waters, as a rule, yield ammonia rapidly, whereas non-decomposing material yields it but slowly, and he points out the importance, therefore, of noting the rate at which the ammonia collects in the distillate.

Dr. Smart has observed that water containing fermenting vegetable matter is colored yellow by boiling with sodium carbonate, and that when Nessler reagent is added to the distillate, a greenish, in place of the ordinary yellowish-brown color is produced. He applies this fact in conjunction with the determination of the oxygen-consuming power (Tidy's process) and the rate of evolution of the ammonia by alkaline permanganate as follows:—

A water yielding ammonia slowly by alkaline permanganate, contains recent organic matter; of animal derivation, if the oxygen-consuming power is low; of vegetable, if high.

A water yielding ammonia more rapidly by alkaline permanganate, shows decomposing organic matter; of animal origin, if the oxygen-consuming power be low and there be no interference with the Nessler reaction; of vegetable origin, if the oxygen consumed be high, and if a yellow color be produced in the water by sodium carbonate, and a greenish color in the nesslerized distillate.

Inferences as to the source of the organic matter can usually be drawn from the amount of chlorin and nitrates present. If the chlorin be high, *i. e.*, in excess of the average of the district, it may be inferred that the material is, in great part, of animal origin. In this case the nitrates will either be high or entirely absent, according as the contaminating matter has passed through soil or enters the water directly.

A large amount of vegetable matter will, as a rule, show itself by the color it imparts to the water.

Wanklyn gives the following standards:—

High purity, ...00 to .041 per million.
Satisfactory purity, ...041 to .082 " "
Impure, ... over .082.

In the absence of ammonium compounds, he does not condemn a water unless the nitrogen by permanganate exceeds .081 per million; but a water yielding 0.123 parts per million of nitrogen by permanganate he condemns under all circumstances.

Total Nitrogen.—Drown and Martin's results with surface waters indicate that the total nitrogen obtained by their process is about twice that obtained by alkaline permanganate. The experiments made by Dr. Beam and myself accord with this. Further observation on different waters and by different observers will be required to determine the value to be assigned to the figures obtained by this method. This method is especially suitable for studying the effects of filtration, storage, etc., on the nitrogenous organic matter in water.

Nitrogen as Nitrites.—Nitrites are present in water as the result either of incomplete nitrification of ammonium, or the reduction of already formed nitrates, under the influence of reducing agents or microbes. Since they are transition products, their presence in water is usually evidence of existing fermentative changes, and, further, may be taken as indicating that the water is unable to dispose of the organic contamination. When, however, the conditions are such that oxidation cannot take place, nitrites may persist for a long time. This sometimes occurs in deep waters in which fermentative changes have long since ceased, but oxygen is not available. These contain not infrequently small amounts of nitrites, to which the same degree of suspicion cannot be attached. When nitrites are found in these waters, the possibility of their introduction from polluted subsoil water, through defective tubing, must not be overlooked. Rain water, also, sometimes contains nitrites derived from the air, and therefore not indicative of any putrefactive change. The presence of measurable quantities of nitrites in river or subsoil water is sufficient ground for condemnation.

Nitrogen as Nitrates.—Nitrates are the final point in the oxidation of nitrogenous organic matter, especially animal matters. Rain water and that from mountain streams and deep wells, except from cretaceous strata, generally contain only traces, but river and subsoil waters will always contain appreciable amounts, unless some reducing action, such as recent sewage-pollution, is at work. When, therefore, a water contains enough mineral matter to demonstrate its percolation through soil, and at the same time is free from nitrates or contains only traces, the occurrence of a destructive fermentation may be inferred.

These cases are not uncommon among well-waters, and the samples are generally turbid from suspended organic matter. Decided departure, either by increase or decrease, from the proportion of nitrates usual in the same class of water in any district may be taken as evidence of contamination.

Oxygen-consuming Power.—Sanitary authorities differ very much as to the significance of this datum. Attempts have been made to fix maximum limits for the various types of water, and also to gauge the character and condition of the organic matter by observing the rate at which the oxidation takes place, but no positive conclusions can be given. In general, it may be said that a sample which has high oxygen-consuming power will be more likely to be unwholesome than one which is low in this respect; but the interferences are so numerous, and the susceptibility to oxidation of different organic matters of even the same type, is so different, that the method is at best only of accessory value. It is especially suitable for consecutive determinations on the same supply.

The following proportions are given by Frankland and Tidy as the basis of interpreting the results of this method:—

Oxygen Absorbed in Three Hours.

High organic purity,	.05	parts per million.
Medium purity,	0.5 to 1.5	" " "
Doubtful,	1.5 to 2.1	" " "
Impure,	over 2.1	" " "

For the method with acidified permanganate at the boiling heat, the German chemists, who employ it largely, regard an absorption of 2.5 parts of oxygen per million as suspicious, and some sanitary authorities have fixed

3.8 parts of oxygen per million as the highest permissible limit.

Dissolved Oxygen.—Full aëration of water is favorable to the destruction of organic matter; a decided diminution in the quantity of dissolved oxygen may show excess of such matter and of microbic life. Gérardin has pointed out that this diminution is associated with the development of low forms of vegetable life, and Leeds has recorded similar facts. These changes are more likely to take place in still waters, and are frequently accompanied by disagreeable odor and taste. In cases in which stored waters become unpalatable, these facts should be borne in mind.

Dupré has given the following as the basis for interpreting the results of his adaptation of the determination of dissolved oxygen:—

"A water which does not diminish in its degree of aëration during a given period, may or may not contain organic matter, but presumably does not contain growing organisms. Such organic matter as it may be found to contain by chemic analysis need not be considered as dangerous impurity."

"A water which by itself, or after the addition of gelatin or other appropriate cultivating matter, consumes oxygen from the dissolved air, at lower temperatures, but does not consume any after heating for, say, three hours at 140° F., may be regarded as having contained living organisms, but none of a kind able to survive exposure to that temperature."

"A water which by itself, or after addition of gelatin or the like, continues to absorb oxygen from the contained air after heating to 140° F., may be taken as containing spores or germs able to survive that temperature."

Hardness.—The degree of hardness has but little bearing on the sanitary value of water, but is important in reference to its use for general household purposes, in view of the soap-destroying power which hard waters possess.

USUAL ANALYTIC RESULTS FROM UNCONTAMINATED WATERS.
Milligrams per Liter.

	Rain.	Surface.	Subsoil.	Deep.
Total solids,	5 to 20	15 upward	30 upward	45 upward
Chlorin,	Traces to 1	1 to 10	2 to 12	Traces to large quantity
Nitrogen by permanganate,	.08 to .20	.05 to .15	.05 to .10	.03 to .10
" as NH_4,	.20 to .50	.00 to .03	.00 to .03	Generally high
" " nitrites,	None or traces.	None	None	None or traces
" " nitrates,	Traces	.75 to 1.25	1.5 to 5	.00 to 3

Inferences from Culture-Methods.—Owing to the great differences in the conditions and manner of growth in various species of microbes, the inferences to be drawn from the cultivation of the germs present in a water-sample are uncertain. In spite of the study which has been given to the subject, the differentiation of specific forms has not reached exactness, nor have the conditions most favorable to growth been ascertained. Different observers pursue different methods of culture, and the results are not comparable. Broadly speaking, the examination of the bacterial life in water may include two inquiries,—a simple calculation of the number of individual living microbes, and the detection of the presence of certain specific forms. Even the first of these objects, though simple, cannot be carried out with absolute certainty at the present day. The number of microbes in water is subject to rapid increase for a brief period after collection of a sample, and may be greatly modified by incidental conditions during storage or transportation, so that little value can be attached to quantitive determination, except when made promptly. The culture-fluids ordinarily used, and the conditions in which the culture takes place, do not suffice for the development of all the forms present. The cultivation ought to be extended over many days, and samples of the water tried with various nutritive media and at different temperatures to secure a knowledge of the forms present. As an indication of the insufficiency of the common methods, it may be mentioned that Miller describes six species of microbes occurring in the human mouth, none of which would grow on any form of culture-medium which he was able to produce.

It is necessary that any record of bacteriologic examination of water should include a precise statement of the conditions of cultivation. That is, composition of the culture-medium; if alkaline or acid, the degree of this reaction expressed in terms of some standard solution; the temperature at which the cultivation takes place, and the time. It is not correct to record the number of individual colonies observed on the culture-field as indicative of so many living microbes, for in some cases two or more microbes may have been jointly concerned in the formation of a colony. A proper method is simply to enter the observation as so many points of microbic life, and to add whatever detailed description is necessary. Efforts have lately been made to secure a consensus of opinion among bacteriologists as to the most satisfactory method of microbe-counting, and to adopt this as a uniform system, in order that results in some degree comparable may be obtained by different observers. At a convention held under the auspices of the American Public Health Association in New York City, June 21st and 22d, 1895, the first step in this direction was taken, but the methods are not yet sufficiently developed to be inserted.

In the determination of specific forms far greater difficulty arises, and unfortunately this difficulty concerns especially the great problem which is presented to the sanitary chemist, namely, whether a given water-supply is likely to produce disease. The recognition of so characteristic a form as the comma-bacillus in water is not very difficult, but it is a problem rarely presented to those chemists who are practicing in civilized countries. The history of the attempts to determine bacilli of objectionable character in water has presented a continuous evolution, in which a degree of confidence existing at one time has been overturned by later discoveries. It was first thought that a distinction might be made between ordinary water-bacteria and sewage-bacteria, by reason of the action of the latter upon the gelatin culture-medium. Some bacteria grow without altering the condition of the gelatin visibly; others rapidly digest it and produce a liquid peptone. It was supposed that these latter were the more objectionable forms, and some chemists still record the liquefying microbes as " sewage microbes." This distinction is, however, untenable. Several bacilli known to be associated with intestinal discharges do not produce any liquefaction, while some that are harmless rapidly liquefy.

The method used in the Hygienic Laboratory of the University of Michigan has the sanction of high authority and the benefit of extended experience, but it is doubtful if it meets the requirements of the sanitary chemist.

However general the acceptance of the method by physiologists, it cannot be considered safe to infer that germs which are poisonous when injected subcutaneously or intra-abdominally into rats, guinea-pigs, rabbits, and mice, are necessarily dangerous in water consumed in the ordinary way by human beings. Tests of this kind require the very highest skill and facilities, involve great expense and delay; and it is probable that in a large majority of cases a careful analysis upon the lines generally accepted will afford as safe a ground for inference as this elaborate method. It is not impossible that treatment at blood-heat and in a highly stimulating nutritive medium may materially increase the virulence of some microbes.

There is, however, one field of inquiry in which even mere microbe-counting has value, and that is in comparing samples of the same water before and after some treatment or other incident. In these studies the method is sufficiently free from fallacy to make the results trustworthy when they are conducted in a strictly uniform manner; thus, if a river-water supplied to a filter be studied daily by examination of repeated samples before and after filtration, inoculating separate portions of the same culture-medium, and multiplying the results to such an extent as to eliminate accidental differences, a comparison between the water before and after filtration may be safely made as to the proportion of microbes removed. Moreover, special microbes of highly characteristic properties may be introduced in large quantities into the water, and by subsequent culture the extent to which these are removed may be satisfactorily recognized. Thus in the extended experiments of the Massachusetts State Board of Health the *Bacillus prodigiosus* was employed as a test-microbe, its chromogenic power enabling it to be detected with great facility. It is said in these reports that the life-history and habits of this microbe are so nearly identical with that of the *B. typhosus* that it is conveniently substituted for the latter in such test experiments. Under the *B. typhosus* the Massachusetts observers refer to a microbe which is obtained from blood and viscera of patients dead of typical typhoid fever. Concerning the alleged specific nature of this microbe, *B. typhosus*, or *B. typhi-abdominalis*, as it has been called (the latter an unfortunate term because it perpetuates a highly inappropriate name for typhoid-fever), it must be noted that there is no satisfactory evidence as to the existence of a distinct, differentiated, specific form, which is the sole and only cause of the fever. The old method of recognizing the *B. typhosus*, by cultivation on potato, was never regarded as very satisfactory, and is now known to be valueless. Research has indicated that several common water-bacteria may assume temporary conditions, known as " involution-forms," in which they may have functions different from those ordinarily belonging to them. Among the bacilli that seem to be closely associated with those forms that have been designated as *B. typhosus*, is one which is extensively present in the intestinal contents of man and the domestic animals, and which is designated as the *B. coli-communis*. It is abundantly present in water which has received any form of sewage, even when that is merely surface-washings. It is not necessarily associated with danger to those drinking the water, but its presence must be regarded as suspicious. It has undoubtedly been designated as *B. typhosus* in many cases, and we now know enough of the errors of this class of investigation to say positively that a large proportion of the earlier literature on this subject is without value. Since it appears that *B. coli-communis* is the germ most likely to be confounded with the specific typhoid-causing germ, if such exists, the efforts of observers have been directed toward indicating the specific differences as clearly as possible. Unfortunately, these differences appear to be entirely negative as to the more important form, rendering its detection in the presence of the more positive form impossible. Thus it is said that the *B. coli-communis* does not curdle milk, while *B. typhosus* does. The former is said to induce fermentation in sugar-solutions, and to furnish the indol-reaction in beef-peptone, while the latter produces neither of these effects. A French observer, however, claims that by growing the *B. typhosus* under new conditions the reactions of the *B. coli-communis* may be produced. A competent American writer has recently stated that "between what we may regard as typical forms of the *B. coli-communis*, on the one hand, and typical forms of the *B. typhosus*, on the other, there is a whole series of intestinal bacilli known under various names and described by various bacteriologists, representing a perfect gradation from one microbe to the other." Klein states that the true typhoid-germ produces an iridescent film on solid gelatin at ordinary temperatures after a few days' growth, and that a precipitate is formed in melted gelatin at 37° C., neither of which actions is exhibited by *B. coli-communis*.

Another water-microbe that has ordinarily no significance of danger is the *B. fluorescens liquefaciens*, but the observations of Moore (Bureau of Animal Industry, No. 3), show that cultures of this microbe suddenly take on septic properties and lose them again subsequently.

Vaughan (*Amer. Jour. Med. Sci.*, August, 1892, p. 198), has expressed himself as follows in regard to the specificity of *B. typhosus* :—

"Of one thing I am certain, and that is, that I am ignorant of any crucial test, or of any combination of tests, upon the strength of which I can say at present that a germ which I may find in drinking-water is identical with the so-called typhoid bacillus. I have found in spleens, after death from typhoid fever, germs which differ from the typhoid bacillus obtained from Berlin, and from one another as markedly as my *B. venenosus* differs from either or both."

ACTION OF WATER ON LEAD.

The almost universal use of lead pipes for conveying water, and the facility with which some waters corrode and dissolve the metal, make it a question of moment to determine the cause of this action and to devise means for its prevention. The subject has received considerable attention within the last few years, and the conditions which determine corrosion are now fairly understood. As a rule, it is found that waters free from mineral matter dissolve lead with facility, especially in the presence of oxygen. Some very soft waters are entirely without action, and this was unexplained until a few years ago, when Messrs. Crookes, Odling, and Tidy found that the action was controlled by the amount of silica contained in the water. They found that those soft waters which, when taken from the service pipes, contained a notable quantity of lead, gave, on the average, three parts of silica per million; in those in which there was no lead, the silica present amounted to 7.5 per million, and in those in which the action was intermediate, 5.5 parts per million. That it was really the silica that conditioned the corrosion, was confirmed by laboratory experiments. They also found that the most effective way of silicating a water is by passing it over a mixture of flint and limestone. The reason for this was pointed out later by Messrs. Carnelly and Frew, who showed that while calcium carbonate and silica both exert a protective influence, calcium silicate is more effective than either, and, further, that in almost all cases in which corrosion took place it was greater in the presence of oxygen. This is particularly the case with potassium and ammonium nitrates and with calcium hydroxid. The reverse is true of calcium sulphate, which is more corrosive when air is excluded. Their experiments also show that the presence of calcium carbonate or calcium silicate, altogether prevents corrosion by potassium and ammonium nitrates.

As the result of an elaborate series of experiments, Müller concludes, that while chlorids, nitrates, and sulphates all act upon lead pipes, no corrosion takes place in the presence of sodium acid carbonate, and that calcium carbonate, by taking up carbonic acid, acts in the same way. This latter conclusion is at variance with the observations of Carnelly and Frew, who found that calcium carbonate is equally effective when carbonic acid is excluded. Müller also states that surface waters, contaminated by sewage and containing large amounts of ammoniacal compounds, will dissolve lead under all circumstances.

Allen has shown that water containing free acid, including sulphuric acid, acts energetically upon lead. This is not surprising in view of the later experiments, which prove that even calcium sulphate is corrosive. Later, W.

Carleton-Williams found that even in the presence of free acid, corrosion may be prevented by the addition of sufficient silica. His experiments also confirm the view generally held, that soluble phosphates protect lead to a marked degree.

The following is a summary of the more important observations on this subject:—

Corrosive: Free acid or alkalies, oxygen, nitrates, particularly potassium and ammonium nitrates, chlorids, and sulphates.

Non-corrosive and preventing corrosion by the above: Calcium carbonate, sodium acid carbonate, ammonium carbonate, calcium silicate, silica, and soluble phosphates.

It is said that filtration through animal charcoal is a means of removing the greater portion of any lead suspended or dissolved in the water. Such filters must be attended to and renovated from time to time.

TECHNIC APPLICATIONS.

Boiler Waters.—The main conditions affecting the value of a water for steam-making purposes are its tendency to cause corrosion and the formation of scale. *Corrosion* may be due to the water itself, to the presence of free acids, or to substances which form acids under the influence of the heat to which the water is subjected. Pure water, *e. g.*, distilled water, exhibits a powerfully corrosive action upon iron. The dissolved oxygen which all waters contain also aids in the corrosion, and especially when accompanied, as is usually the case, by carbonic acid. There is always greater rusting at the point at which the water enters the boiler, since there the gases are driven out of solution and immediately attack the metal.

This is an evil that obtains with all waters, and it is not customary, in making examination for technical purposes, to determine the amount of these bodies. In water that has had free access to air, the oxygen in solution is a tolerably constant quantity, and it is sufficient to note the temperature and refer to the table of amounts of oxygen dissolved in water. The corrosive action of oxygen and carbonic acid is especially noticeable in waters that are comparatively pure, such as those derived from mountain springs. This was repeatedly observed by Dr. William Beam, in the examination of the waters used for the locomotives of the Baltimore and Ohio Railroad. The waters which caused the most corrosion were mainly those containing small quantities of solid matter, the full amount of oxygen and considerable carbonic acid, but no other acid or acid-forming body.

Free acid, other than carbonic acid, is not often found in water, and if present renders the water unfit for use, unless it be neutralized. Mine waters are the most likely to contain free acid, sulphuric acid being generally present. Sometimes the acidity is due to organic acids. These act very injuriously on iron. Allen gives an example of this in the water supplied to Sheffield, Eng., which he found to contain an organic acid in amount equivalent to from 3.5 to 10 parts of sulphuric acid per million.

Magnesium chlorid is frequently present in waters, and if in considerable quantity may be very harmful. At a temperature of 310° F., corresponding to an effective pressure of four atmospheres, magnesium chlorid reacts with water to form magnesium oxid and hydrochloric acid, the latter attacking the boiler, especially at the water line.

If there is present at the same time considerable calcium carbonate the evil may be somewhat lessened, but as Allen has pointed out, and as we also have noticed, there may still be corrosion, so that the presence of more than a small quantity of the salt, say a grain or two to the gallon, may be considered objectionable. Allen remarks that the presence of a certain amount of sodium chlorid may prevent this decomposition, the two chlorids combining to form a stable double salt. The addition, therefore, of common salt to a water containing magnesium chlorid may act to diminish corrosion, a point which will bear further investigation.

It has not been determined how far the presence of nitrites, nitrates, and ammonia affects the quality of water for steam-making purposes; but it is more than probable that they act harmfully, especially the nitrates, which are frequently present in large amount.

Scale is composed of matters deposited from the water either by the decompositions induced by the heat or by concentration. When the deposit is loose it is termed *sludge* or *mud*, and usually consists of calcium carbonate, magnesium oxid and a small amount of magnesium carbonate. The magnesium oxid is formed by the decomposition of the magnesium carbonate and chlorid. This fact was first pointed out by Driffield (*J. Soc. Chem. Ind.*, VI, 178).

The formation of sludge is the least objectionable effect, since it may readily be removed by "blowing off," provided that care is previously taken to allow the flues to cool down so that when the water is removed the heat of the flues may not bake the deposit to a hard mass. Waters containing calcium sulphate form hard incrustations difficult to remove and causing great loss of fuel by interfering with the transmission of the heat to the water. It not only forms a hard incrustation in itself, but becomes incorporated with the mud, and renders it also hard. The hard scale will also contain practically all the silica and the iron and aluminum present in the water, besides any matters originally held in suspension.

It follows from the above that a water only temporarily hard will, if care is taken in the management of the boiler, cause the formation merely of a loose deposit of sludge—temporary hardness being due in the main to calcium and magnesium carbonates. A water permanently hard will probably form a hard scale, since such hardness is usually due to calcium sulphate.

In accordance with these principles, the analysis of a water for steam-making purposes may include the determinations of free acid, total solid residue, SO_4, Cl, Ca, Mg, temporary and permanent hardness. In cases in which the qualitative tests show but small amounts of SO_4 and Cl, the analysis may be limited to the determinations of the temporary and permanent hardness.

It has been pointed out in an earlier chapter that it is not possible to deduce from the analytic result the exact forms in which the various elements are combined, but since it is known that at the high temperature ordinarily reached in boilers definite chemical changes occur, it is safest to exhibit the maximum amount of corrosive and scale-forming ingredients which the water under these circumstances could develop. Thus, since calcium sulphate is practically insoluble in water above 212° F., the proportion of calcium sulphate may be regarded as such as would be formed by the total quantity of calcium or the total quantity of SO_4, according to which is present in the larger amount. Similarly, as the decomposition of magnesium chlorid is induced by the high temperature of the boiler, the analytic statement should indicate the maximum proportion of this compound obtainable from the magnesium and chlorin present. These rules cannot apply absolutely to waters rich in alkali-carbonates, since these would neutralize any acid formed from the magnesium chlorid, or even prevent its formation, and would prevent to a large extent the formation of calcium sulphate. Much remains to be determined concerning the effects of the high temperature and concentration to which boiler waters are subjected.

General Technic Uses.—In regard to the quality of water for technic other than steam-making purposes, such as brewing, dyeing, tanning, etc., no detailed methods or standards can be laid down. The nearest approach to purity that can be secured in the supply will be of the greatest advantage. The more objectionable qualities will be large proportion of organic matter, especially if it distinctly colors the water, excessive hardness, and notable amounts of iron or free mineral acid. It is stated by Bell (*Jour. Soc. Chem. Ind.*) that one part per million of iron will render water unsuitable for bleaching establishments. It has been noted that a large proportion of active microbes is injurious in the manufacture of indigo. In artificial ice making, a very pure water must be used if a clear and colorless product be desired. Any suspended or dissolved coloring matter will be concentrated by the freezing and appear in the bottom or center of the mass. The Antwerp water, purified by the Anderson process, is used with entire satisfaction for the manufacture of artificial ice in that city.

The examination of sewage-effluents and waste waters from manufacturing establishments is to be conducted upon the same principles as for ordinary supplies, but especial attention must be given to the presence of poisonous metals, and free mineral acids. The latter interfere with the normal self-purification of the water. For the nitrogen determination, the Kjeldahl process will be found more satisfactory than that by alkaline permanganate.

PURIFICATION OF DRINKING-WATER.

The most obvious method of purifying water is by distillation. The process is too expensive for general use, but is especially adapted for water intended for pharmaceutical or chemical purposes. It has also been used for supplying vessels at sea and in tropic localities in which the natural waters may be contaminated with malarial or other germs. The majority of microbes are killed by short exposure to a temperature of 212° F.; hence, water may be purified, on a small scale, by simple boiling. Freezing does not have the same effect, many microbes retaining vitality for a long while in ice.

The self-purification of water, that is the destruction of organic matter and pathogenic microbes, by reason of the development of the ordinary microbes of putrefaction, occurs satisfactorily only in alkaline waters, hence, acid effluents check this process. The addition of lime in sufficient amount to give a slightly alkaline reaction will be beneficial.

The conditions necessary to secure self-purification of

surface waters are not fully understood, and it is unsafe to state that a polluted stream will purify itself in any given distance. It must also be remembered that the dilution of infected sewage by its introduction into a large volume of uninfected water will assist, for a time, at least, the multiplication of pathogenic microbes.

Heider made a study of the water of the Danube. Although there is not complete commingling for several miles after the sewage inflow, yet, as soon as the sewage is diluted somewhat more than seven times, the analytic differences between the river water above and that below the contaminating inflow are exceedingly small, yet the bacteria remain several times as numerous for miles down stream after the introduction of the sewage-water. Muscle-tissue stained by bile was found after the water had flowed for twenty-five miles. The current of the river is from four to seven miles an hour. Gruber and von Kerner have shown that cholera germs can remain alive for from five to seven days in the river-water, as also in the Vienna aqueduct-water. Hence the self-purification is a matter of dilution, and, as elsewhere, should not be relied upon when epidemics of water-borne infectious diseases are present. Pupils of Pettenkofer who studied the Isar water at Munich, consider that self-purification within twenty miles was there more reliable and certain. Fraenkel has recently found that the Lahn purified itself speedily after the sewage of Marburg entered it.—(*N. Y. Med. Jour.*, Dec. 1, 1894.)

The methods in general use for purifying water are simple filtration and the removal of the impurities by appropriate chemical agents.

Filtration.—For household purposes forms of carbon, stone and sand filters are used, which yield clear filtrates, but permit, sooner or later, the transmission of microbes. The suspended matter in the water gradually accumulates on the surface of the filter, and causes a great increase in the number of the microbes, some species of which apparently grow through the pores of the filter, and are carried into the filtrate. The following are among the more efficient forms of household filters:—

Bischof Spongy-Iron Filter.—The construction of this is shown in Fig. 10. The spongy iron is obtained by reducing hematite, at a temperature below the fusing point of iron.

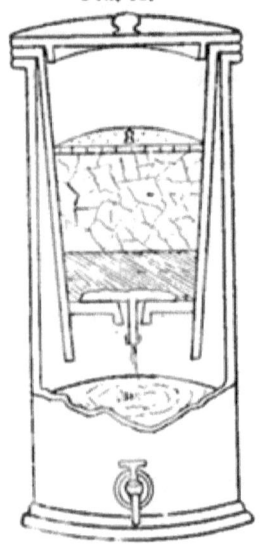

Fig. 10.

It rests on a layer of pyrolusite (manganese dioxid), below which is an asbestos bag having a short tube with perforated cap. This is a very efficient form, removing much of the dissolved organic matter, and practically all the suspended matter, including the microbes.

Chamberlain-Pasteur Filter.— This consists of tubes of unglazed biscuit-ware, the number depending on the size and required delivery of the filter. There are arrangements for continuous filtration by attaching the tube to the faucet; also forms adapted to simultaneous cooling and filtration. The observations of Pasteur and others have shown that this is a highly efficient filter, yielding for a considerable time a filtrate entirely sterile. It requires occasional cleaning, since, after continuous use, the microbes may pass through the pores, probably by a process of growth. An occasional boiling of the tubes in water would be sufficient to overcome this difficulty.

Fig. 11 shows a form of sand filter which is used in the laboratory of Professor Kemna, at Antwerp. A moderately wide and stout tube is passed to the bottom of a tall jar, and the intervening space filled to the depth of about 25 cm. with fine sand, coarse sand, and gravel, as shown. The exit tube consists of a siphon, the outer leg of which does not quite reach to the level of the surface of the sand, the inner leg reaching to the bottom of the jar.

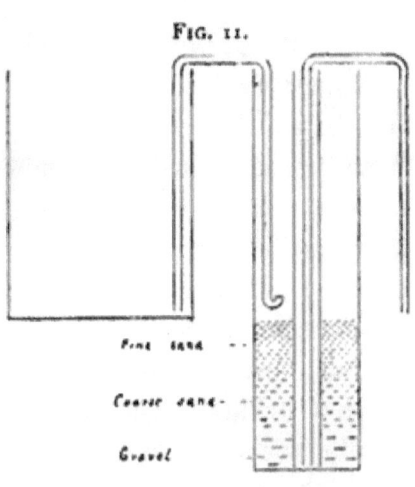

FIG. 11.

The flow may be controlled by a stop-cock attached to the outer leg. The object of this arrangement is to prevent the water-level being drawn to or below the level of the sand. The filter should be supplied from a reservoir by means of a siphon, the exit tube of which is curved upward, in order to prevent disturbing the deposit which collects on the surface of the filter.

The filter may be cleaned by removing the siphon and sending a slow current of water down the wide tube until the deposit upon the surface of the sand is washed out. The apparatus is especially suited for laboratory experiments on filtration.

Many other forms of filter have been devised. Comprehensive comparative tests made under the auspices of the *British Medical Journal* show that, except as to those based on the principle of the Chamberlain-Pasteur filter, but little time elapses before the filtrate contains numerous microbes.

For the purification of drinking-water on a large scale, sand filter-beds have been found to be efficient; but the best results are obtained only under proper supervision.

Numerous determinations of the efficiency of sand filtering basins have been made by various methods. It has been found that, at the start, a large proportion of the organic matter, dead and living, passes through; but that as filtration proceeds, the surface of the sand becomes covered with a close deposit, which acts both as a means of retaining suspended impurities and, by its active microbic life, destroys the organic matter in a manner analogous to that occurring in soil. The water thus becomes practically free from microbes, but after a time these gradually penetrate the pores of the filter and appear in the filtrate. Increase of pressure will hasten this effect.

Bertschinger (*Jour. Soc. Chem. Ind.*, Dec., 1889) has published observations on the efficiency of the sand filters in use at the Zurich Water-works. The filtering material rests on a brick grating, and consists of the following layers, commencing at the bottom:—

Five to 15 cm. of coarse gravel, 10 cm. of garden gravel, 15 cm. of coarse sand and 80 cm. of fine sand. As soon as the diminution in pressure of water, due to the resistance of the filter, is from 60 to 80 cm., the filter is cleaned by allowing the water to run off and removing the top layer of sand to a depth of two cm., as this is found to contain the whole of the mud. The filter is then filled up with filtered water from below, and washed by allowing this to overflow. After filtration has recommenced, the first portion of the filtrate is rejected. Two of the filters are arched over; these require cleaning once in seventy-seven days, the others once in forty-eight days. As soon as the layer of fine sand has been reduced to the thickness of 50 cm., fresh sand is substituted or more added, until the depth is again 80 cm. Among the conclusions reached are the following:—

Under normal conditions the filtered water is free from microbes, although a few are taken up again in the later stages of the filtration.

After cleaning the filter, the water which first passes through is not in normal condition. It contains many microbes, the efficient layer of scum not having had time to collect on the sand, though the chemical purity of the water is satisfactory.

When the filters have not been used for some time, the water which first passes through them contains more bacteria than usual, owing to their rapid multiplication in stagnant water, but its chemical purity is not materially different from the normal filtered water.

Sand-filtration has been brought to practical use at Lawrence, Massachusetts, where a filter has been in operation for several years, with the effect of markedly reducing the death-rate from typhoid-fever in that city. The water supplied is that of the Merrimack River, which ten miles above the Lawrence intake receives the sewage of Lowell, a town of 80,000 population. The following is a description of the Lawrence filter given by Mr. George W. Fuller, biologist in charge of the Lawrence Station. It was designed by Hiram F. Mills, Engineer-member of the Massachusetts State Board of Health.

It is 2.5 acres in area, and contains sand of an average

depth of about 4.5 feet. The depth of sand varies from three to five feet, but owing to the arrangement of the underdrains all water passes through at least five feet of filtering material. The filter is situated by the side of the Merrimack River, and separated from it by an embankment. Its surface is two feet below low water in the river. The water is allowed to flow on to the filter about 16 hours a day on an average, and during the remainder of the time the sand is drained and the pores filled with air. The filtered water is conducted by underdrains to a collecting conduit, and thence to the pump-well. The pumps determine the rate of filtration, and are speeded so that the water shall pass through the filter at the rate of 2,000,000 gallons per acre per day. From the pumps the water passes to the open distributing reservoir, which is 25 feet deep at high water and contains 40,000,000 gallons. The water then flows by gravity from the reservoir to the consumers.

From the time when the filter was put in operation, September 20, 1893, until May 1, 1894, daily bacteriologic tests, in addition to numerous analyses, were made of the water before and after its passage through the filter, as it leaves the reservoir, and from taps at the City Hall and Experiment Station, which are distant 1.5 and 2.5 miles, respectively, from the reservoir. The results were as follows:—

	Average number of bacteria in one c.c.	Percentage removed.
River,	10,900	...
Effluent at Filter,	264	97.58
" " Reservoir Outlet,	130	98.73
" " City Hall,	90	99.17
" " Experiment Station,	82	99.25

The above averages include all results. Excluding those results obtained under conditions which were abnormal and are not likely to occur again, we find that this filter normally reduced the bacteria from 9000 to 150 per cubic centimeter, a removal of 98.3 per cent. of the number applied. Owing to the fact that some ground water of somewhat unsatisfactory quality with regard to numbers of bacteria was at times mixed with the effluent, it is very improbable that all the bacteria in the water pumped to the reservoir passed through the filter.

The following account of some details of the construction of filters for sewage-purification has been furnished me by H. W. Clark, Chemist at the Massachusetts Experiment Station, Lawrence, Mass.: The sands in use in the large filter vary in effective size from .04 mm. to 1.40 mm. That is to say, the finest ten per cent. of the material is composed entirely of grains whose diameter is less than .04 mm. in the finest material used, and less than 1.40 mm. in the coarsest. With coarse and medium-fine sand a filter contains but one grade throughout the entire five feet. With fine sands, trenches are sometimes dug one to two feet deep, and filled with coarser sand. This gives a given area greater filtering capacity. With what is considered the best grade of sand the applied dose of sewage, one hundred thousand gallons per acre per day, passes below the surface of the sand in a time varying from ten minutes to one-and-a-half hours, depending mainly on the condition of the surface of the filter. The remainder of the twenty-four hours the surface is uncovered. In the intermittent filtration of water as now practiced at this station, the surface of the filter is uncovered two hours out of the twenty-four.

The liability to the free passage of microbes for a short time after filter-cleaning has been often alleged as a drawback in the use of sand-filtration. It is thought that the formation of a close film of microbes upon the surface of the filter is necessary to perfect working, but the observations at Lawrence have indicated that this period of diminished efficiency is due in part to the mechanical disturbance of the main body of the filtering material, during the process of refilling with water after the draining and scraping. Slowly filling the filter from below upward has been found to be a reliable method of avoiding the difficulty. It is stated by Mr. Fuller that a much greater speed than 2,000,000 gallons per acre per day can be employed without diminishing the efficiency for at least a time, but it is not yet known how this will affect the permanent efficiency of the filter.

Precipitation Methods.—The observations of Dr. P. F. Frankland and others have established the following points :—

" Organized matter is, to a large and sometimes to a most remarkable extent, removable from water by agitation with suitable solids in a fine state of division, but such methods of purification are unreliable.

" Chemical precipitation is attended with a large reduction in the number of microörganisms present in the waters in which the precipitate is made to form and allowed to subside.

" If subsidence either after agitation or after precipitation be continued too long, the organisms first carried down may again become redistributed throughout the water."

It is essential, therefore, that the liquid be filtered as short a time after the precipitation as possible.

Precipitation by Aluminum Compounds.—A small quantity of aluminum sulphate added to natural waters is decomposed with the formation of a flocculent precipitate of aluminum hydroxid, which settles comparatively rapidly, and carries down with it all suspended matters, as well as a large proportion of the dissolved organic matters. Waters which contain such an excess of organic matter as to be distinctly colored, may usually be made quite clear and colorless by this treatment. One grain to the gallon will suffice for the purpose, but if very rapid subsidence is desired more may be added.

This precipitation is a gradual process, and a water that will give the test for aluminum immediately after filtering may give none after twenty-four hours. It is not infrequently noted that such effluents, originally clear, become cloudy on standing, in consequence of the separation of aluminum hydroxid. On the addition of aluminum sulphate to brown surface-waters there is also a precipitation by the organic matters. A sample of the Cochituate-river water (Boston supply), of moderately deep color, to which 25 milligrams of alum to the liter had been added, when filtered gave no reaction, even when 2.5 liters were concentrated for the test. An addition of 30 milligrams to the liter could be detected without difficulty.

Several systems of filtration now in extended use employ this precipitation method in conjunction with filters of small area, the necessary flow being obtained by increased pressure. The differences between the various forms are chiefly in the mechanical arrangements for supplying the water and for cleaning the filter. The material is generally sand or coke; the cleaning is performed at short intervals, by means of reverse currents of water. The aluminum solution is introduced as needed by automatic apparatus.

These filters are efficient, and are suitable for the purification of water for manufacturing establishments, and when large basins are not available.

The addition of an iron salt to water containing carbonates, is attended with decomposition and the formation of a precipitate of ferric hydroxid. This reaction has been employed with great advantage as a means of purification. One of these methods was by passing the water through spongy iron, then aërating to precipitate the iron, and filtering through sand. The method is very efficient, but the spongy iron gradually chokes by oxidation and becomes useless. This difficulty is removed by the use of iron borings or punchings, contained in an iron cylinder (Anderson and Ogston, *Proc. Inst. Civ. Eng.*, Vol. 81), which is rotated while the water passes through; the iron is brought into thorough contact with the water, and there is sufficient abrasion to keep its surface clean.

The apparatus as practically employed is shown in Fig. 12. It consists of a cylinder rotating in a horizontal position, attached to the internal periphery of which are short curved shelves, arranged at equal distances. Pipes enter the hollow trunnions to admit and discharge the water. As it enters the cylinder, the water strikes against a circular distributing plate, and is caused to flow radially through a narrow annular space, to prevent the formation of a central current along the axis of the purifier. The inner end of the outlet pipe carries an inverted bell-mouth which catches the fine particles of the iron carried forward by the water,

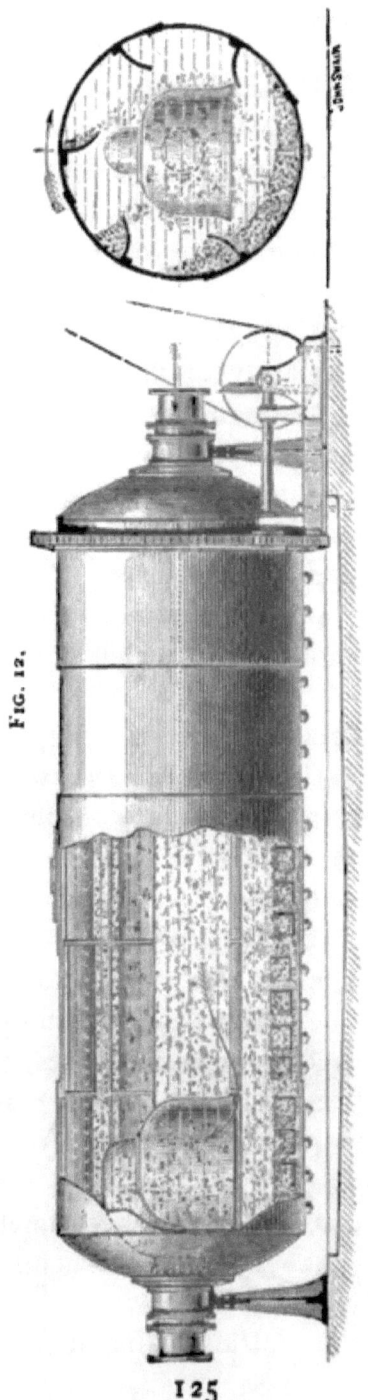

FIG. 12.

125

and causes them to fall again to the bottom of the cylinder. Sufficient borings or punchings to one-tenth fill the cylinder are introduced, and the purifier is then completely filled with water, and set in motion, the rate of rotation being about six feet per minute at the periphery. The effect of the rotation is to scoop up the iron particles and to shower them down through the flowing water.

The effect is due mainly to the formation of ferrous carbonate, through the action of the carbonic acid of the water. On issuing into the open air this is gradually converted by oxidation into the insoluble ferric hydroxid, which carries down much of the organic matter and subsequently oxidizes and destroys it. Temporary hardness is also decreased by the abstraction of the carbonic acid and consequent precipitation of calcium and magnesium carbonates.

The time of contact with the iron depends upon the purity of the water. For Antwerp water, which is purified by this means, the maximum effect is accomplished in 3.5 minutes. After leaving the cylinder, the water is passed through sand filters. Analytic examinations show the effluent water to be of high organic purity and practically sterile.

Cast-iron borings are much more readily acted upon than steel punchings. The latter suffice in operating upon water rich in dissolved organic substances, carbon dioxid and suspended matters, while the former are especially suited for treatment of water comparatively pure, and, therefore, less active.

For determining the most advantageous method of treating any water supply, laboratory experiments may be made as follows:—

A strong wide-mouth bottle holding about 2000 c. c., is

charged with one-tenth its bulk of clean borings or punchings, filled completely with water and shaken for four minutes, in such manner that the iron particles are continuously showered through the liquid. The water is then aërated by agitating it in a large, thoroughly clean glass-stoppered bottle. With some (*e. g.*, peaty) waters, it will be of advantage to allow about 200 c. c. of air to remain in the bottle in which the shaking with iron is performed, occasionally removing the stopper to renew the air. (Continuous aëration during treatment is provided for in the apparatus used on the large scale.) The liquid is allowed to stand from a few minutes to four hours, depending on the rapidity with which the iron separates, and is then filtered through the sand filter (Fig. 11), or through a well-washed cotton plug inserted in the neck of a funnel. Satisfactory purification as regards ammonium compounds cannot be obtained, but with proper attention to cleanliness, the figures for total organic nitrogen, nitrogen by permanganate, and oxygen-consuming power will be trustworthy.

The following results were obtained by Dr. Beam and myself in the treatment of Delaware River water at Lardner's Point pumping station, Philadelphia. The purifier was capable of delivering 100,000 gallons per 24 hours. All figures are in parts per million.

September, 1890.

	Before Treatment.	After Treatment.
Nitrogen as ammonium,	.04	.03
" as permanganate,	.27	.09

November, 1890.

Nitrogen as ammonium,	.084	.034
" by permanganate,	.091	.049
" as nitrites,	traces.	none.

The apparatus was subsequently transferred to Belmont

pumping station, Philadelphia, and bacteriologic examinations made. The applied water was taken from Belmont reservoir which was supplied from the Schuylkill River. The cultivation was made in alkaline meat-extract-peptone-gelatin at ordinary temperatures. The figures represent the number of points of microbic life in one c. c. of the sample.

Date.	Applied water.	Effluent water.
Nov. 11,	innumerable	36
" 13,	4,900	32
" 30,	2,860	49
Dec. 7,	22,400	79
" 11,	17,100	560
" 14,	16,000	60
" 18,	14,500	27

The sample for December 11 was taken just after filter-cleaning. Moreover, the weather had been very cold and a film of ice formed around the sides of the tank and over the bottom; the expansion of this ice pushed the sand away from the sides of the tank. A thaw occurred and a space was left all around the sand-bed through which the water passed without proper filtering.

Laboratory experiment on a sample of water from the Mississippi River at Memphis, Tenn.

	Before Treatment.	After Treatment.
Oxygen absorbed at 212° F.,	2.88	0.36

In the laboratory of the State Board of Health of Massachusetts Dr. Drown investigated the effect of various methods of aëration, such as exposing water in bottles to the air of the room, drawing a current of air through by means of an aspirator, shaking it in a bottle by machinery, and exposing it to air under pressure of from 60 to 75 pounds. While no appreciable benefit as far as regards the organic matter and its decomposition-products occurs, aëration appears to prevent the growth of algæ, with the troublesome accompaniments of bad tastes and odors. It may also have a beneficial effect upon ground-waters containing considerable amounts of iron. These waters are often clear when first drawn, but become turbid and yellow in a few hours by the separation of ferric-hydroxid by oxidation. Waters from considerable depth often contain so little free oxygen that this oxidation does not occur until they reach the surface. By applying an aëration method the change may be hastened, and by some simple process of rapid filtration afterward applied, the water will be made clear and remain so.

A material, designated by the patentee as "magnetic carbid of iron," has been used for the purification of sewage. It is made by heating hematite with coke and sawdust, so as to reduce the ore to the composition of magnetic oxid.

Purification of Boiler Waters.—The problems present in the treatment of boiler waters are usually the removal of the calcium carbonate and sulphate, and magnesium carbonate and chlorid. Both carbonates are appreciably soluble in pure water. About one grain of calcium carbonate to the gallon is usually stated to be the proportion dissolved, but it has been pointed out by Allen that solutions can be obtained containing twice this amount. If the water contains carbonic acid it will take up a much greater proportion of the carbonates, but in this case they will be deposited from the solution by boiling. This has been accounted for by supposing the existence of soluble bicarbonates, which are decomposed by the boiling.

Nearly all of these carbonates can be thrown out of solution by any means that will deprive the water of the carbonic acid. Sodium hydroxid is often employed for the purpose, and should be added in quantity just sufficient to form normal sodium carbonate. If there are present in the water calcium and magnesium chlorids and sulphates, these also will be decomposed and precipitated by the sodium carbonate so formed. If the amount of sodium carbonate formed is not sufficient to decompose all of these bodies, a sufficient quantity should be added with the sodium hydroxid to effect the complete decomposition. The precipitate is allowed to settle or filtered off.

In cases in which the feed-water is heated before it enters the boiler, it may only be necessary to add to the water sodium carbonate in quantity sufficient to decompose the calcium and magnesium chlorids and sulphates, since the heat alone will suffice to throw down the carbonates.

Care should be taken in these precipitations that no more sodium hydroxid is added than is required for the precipitation, since any excess would tend to corrode the boiler.

Clark's process consists in treating the water with calcium hydroxid (lime-water). This precipitates the calcium and magnesium carbonates by depriving the water of its free carbonic acid. It has, of course, no effect upon the calcium sulphate. It is to be noted that the proportion of calcium hydroxid which is to be added must be calculated from the amount of free carbonic acid existing in the water, and not from the amount of carbonates to be removed. The precipitate will usually require at least twelve hours for complete subsidence, but after three or four hours the water will be sufficiently clear for some purposes. If a filter press is used, as in Porter's process, the time required for clarification is very much shortened. Another advantage of this process is the use of a solution of silver nitrate, in order to determine more conveniently the proportion of calcium hydroxid which is to be employed. The lime is first slaked and dissolved in water, and the water to be softened run in and thoroughly mixed with it. From time to time small portions are taken out and a few drops of a solution of silver nitrate added. As long as the lime is in excess a brownish coloration is produced. When this has become quite faint, and just about to disappear, the addition of the water is discontinued, and, after a short time, the water is filtered by means of the press.

Soluble phosphates added to a water, precipitate completely in a flocculent condition any calcium, magnesium, iron, or aluminum. This reaction can be best applied by using the tri-sodium phosphate ($Na_3PO_4 + 12H_2O$), which is now a commercial article. By reason of the facility with which this substance loses a portion of its sodium to acids, it acts not only as a precipitant to the above materials, but will neutralize any free mineral acid present in the water. From evidence submitted by those who have used the process on the large scale, it appears that not only is no hard scale formed, but that scale already existing prior to its use is gradually disintegrated and removed with the sludge. Experiments indicate that no injury results from an excess of the material; but the economical employment of the method, especially with very hard waters, can only be based upon a correct analysis, and an estimation of the phosphate required for the precipitation. In many cases the composition of the water will be such that a partial precipitation will be sufficient.

Of late years considerable success has been obtained by the use of fluorids as precipitants of scale-forming elements.

Waters rich in ferrous compounds may be purified by thorough aëration and filtration, the iron being separated as ferric hydroxid. Simple filtration through a bed of manganese dioxid will accomplish the same purpose.

IDENTIFICATION OF THE SOURCE OF WATER.

The determination of the course of underground streams, and of communications between collections of water, is often an important practical problem. In geologic and sanitary surveys, valuable information may occasionally be gained. The method generally pursued when connection between water at accessible points is to be detected, is to introduce at one point some substance not naturally existing in the water, and capable of recognition in small amount. Lithium compounds are among the best for this purpose. They are not frequent ingredients of natural waters, and are easily recognized by the spectroscope. Lithium chlorid is the most suitable. The quantity to be employed will vary with circumstances. It scarcely needs to be stated that the waters under examination should be carefully tested for lithium before using the method.

When the lithium method is inadmissible, recourse must be had to other substances of distinct character, such as strontium chlorid, but this possesses the disadvantage that a considerable amount may be rendered insoluble, and thus lost in the ordinary transit through soil. Recently, use has been made of organic coloring matters of high tinctorial power, one of the most suitable of which is *fluorescein*, $C_{20}H_{12}O_5$, a derivative of benzene. This will communicate a characteristic and intense fluorescence to many thousand times its weight of water. An entire river may be colored by a few kilograms. By its use an underground communication was demonstrated to exist between the Danube and the Ach, a small river which flows into the Lake of Constance. The coloration is distinct only in alkaline liquids. Other colors, such as anilin-red, may be employed. For detecting leakage from cesspools and cisterns, sanitary inspectors occasionally employ water colored by Prussian blue.

I am indebted to Dr. F. P. Vandenburgh, who conducted the investigation, for a description of an instance of the application of the above methods. In a suit at law growing out of use of a creek for the supply of Syracuse, N. Y., it was alleged that the creek supplied a spring which was used by a manufacturing establishment. Tests of the water of creek and spring for lithium were made, ten gallons of each being evaporated, with negative results. Twenty-five pounds of lithium carbonate were converted into chlorid and poured into the stream about half a mile from the spring. Samples of ten gallons each were taken out of the spring by almost continual dipping during forty-eight hours following. Twenty of these samples were examined and lithium found in each.

Ten pounds of fluoresceïn were introduced at a point about one mile above the spring and the characteristic fluorescence appeared at the spring about six hours after its introduction into the creek. The greatest intensity of color was between six and ten hours after its introduction.

A more important feature of the problem in a sanitary point of view is the determination of the source of a given current or collection of water, when such source is inaccessible. Problems of this character are not infrequent in large cities in which the systems of water supply and drainage are defective, thus giving occasion to accumulations of water in cellars and similar places. Often, in these cases, no extended explorations can be made, by reason of the adjacent buildings and conflicting property interests, and the question may arise whether the water proceeds from a leaky hydrant, drain, sewer, or subsoil current. It is obvious that in the case of the collection of water in a cellar from causes other than surface washings or entrance of rain, it must have passed through some distance of soil, and in built-up districts will almost certainly be charged with organic refuse. To correctly interpret the results, it will be necessary to know the general character of the subsoil water of the district and the composition of the public supply. As a rule, the transmission of water through moderate distances of soil will not materially increase the mineral constituents. Hence, if the sample contains an excess of dissolved matters as compared with the water supply of the district, it may reasonably be inferred that it is derived from a drain, sewer, or subsoil current.

In these investigations it will generally be sufficient to determine the total solids, odor on heating, chlorin, nitrates and nitrites. The following figures are from some results obtained in investigations made in association with Mr. Chas. F. Kennedy, Chief Inspector to the Board of Health of this city :—

	City Supply.	Cellar Water.		
		No. 1.	No. 2.	No. 3.
Total solids,	115	140	661	640
Odor on heating,	faint	faint	strong	urinous
Chlorin,	4	6.4	77.0	128.0
N as nitrates,	0.7	1.0	3.5	none
" " nitrites,	none	present	present	none

Sample No. 1 was taken from a cellar in which a small amount of water had been almost constantly present for a long time, and of which the source could not be ascertained. The results of analysis led to the view that since it resembled in composition the city supply, it was derived from a leaky hydrant pipe. The parties in interest were not inclined to accept this opinion, but the examination of the condition of the hydrant on an adjacent property showed a leak, which being repaired the water ceased to appear in the cellar. In this case it was found that the water had passed through twenty-two feet of earth. In the second case the sample is seen to be very impure, and it was suggested that it was derived directly from a leaky drain, which upon exploration proved to be the case. In the third sample, the high chlorin, strong urinous odor and absence of nitrates and nitrites, pointed unmistakably to recent and profuse contamination with sewer water.

Occasionally the analytic results will be ambiguous, and it is advisable to make examinations of more than one sample, since accidental circumstances, rain-fall, etc., may affect the composition of the water.

Instances of the contamination of water by unusual substances are occasionally noted, and these sometimes afford a clue to the source of the water. Among the instances of this kind within my own experience may be noted the contamination with petroleum and with soap. In the former case it was evident that the contamination was from a leaky pipe connecting two refineries. In the latter it was shown to be derived from an adjoining building used as a laundry.

ADDENDA.

To page 63.

Extended observations, especially in Massachusetts, have shown that reservoirs intended for even moderately prolonged storage of water should be clean, that is, organic matter of any kind should not be allowed to accumulate on the bottom and sides. Dr. Drown states (*Rep. S. B. of H. of Mass.*, 1891) that while the water in one basin became foul from stagnation, in another which was carefully prepared by the removal of all soil and vegetable matter, and is supplied by a brown, swampy water from a district almost entirely free from pollution, the water is good at a depth of forty feet.

In Philadelphia, where large storage reservoirs are used for water that is often very muddy, but little trouble from the growth of microscopic organisms occurs. These reservoirs are artificial basins.

To page 64.

Owing to the great differences in the size of microscopic organisms, the mere enumeration of their numbers is not always an index of the amount of living matter in suspension. To obviate this, Mr. Geo. C. Whipple, biologist to the Boston Water Board, has suggested a standard unit of size, estimating by means of it the total volume of the organisms, and not their number. He finds by this method that the analytic and biologic results correspond much more closely than when mere numbers are recorded. The unit is an area of 400 microns, that is, a square of 20 microns on a side. The results are stated in number of standard units per c. c. A detailed account of the method and of some results obtained by it will be found in *The Amer. Month. Mic. Jour.*, 1894, p. 337.

Mr. Whipple has investigated the conditions influencing the growth of the microscopic organisms in water. He finds that diatoms thrive best with a supply of nitrates and a free circulation of air; temperature alone has no very direct effect. Infusoria will be found in largest numbers when the water contains the greatest amount of finely divided organic matter. When the conditions bring about a circulation of the water, the organisms are not only brought constantly in contact with new food materials, but are enabled to reach the upper layers of the water where oxygen is abundant.

To page 65.

In a private communication of recent date, from Mr. George W. Fuller, of the Experiment Station at Lawrence, Massachusetts, I am furnished with the following information of great practical value in the bacteriologic examination of drinking water. At this station the methods for the isolation of the so-called typhoid bacillus are those described in the "Report of the Massachusetts State Board of Health" for 1891, page 637. The only additional step is to employ Wurtz' litmus-lactose-agar. This consists of nutrient agar-agar of definite reaction. See below for the methods for determining reaction, to which 2 to 3 per cent. of lactose is added. It is sterilized and then enough sterilized litmus solution added to give a decided blue tint. The bacilli included under the designation *B. typhus abdominalis* do not produce fermentation with this medium, but some common water-bacteria do. (Abbott, "Manual of Bacteriology.") This is found to be of much assistance.

The methods are employed with reasonable confidence at Lawrence, where the observers are fairly familiar with the bacterial flora of the water examined. Concerning waters from elsewhere much significance is attached to the presence or absence of the *B. coli communis*.

Mr. Fuller has also favored me with a comprehensive abstract of a valuable communication presented by him to the Convention of Bacteriologists, to which reference is made on page 103. After discussing the influence of reaction upon the various phenomena taking place in culture-media, it was pointed out that all regular culture-media as used are, strictly, neither neutral, acid, or alkaline, but the reaction depends upon the indicator used. Many are alkaline to litmus and acid to phenolphthalein; the cause of this is twofold,—the presence of acid phosphates and of proteids; the latter are amphoteric, but the acid reaction predominates. Phenolphthalein is the best and most delicate agent for determining reaction in culture-media. Under ordinary circumstances the phosphate present is Na_2HPO_4. To all other indicators, except turmeric and phenolphthalein, this salt is alkaline, to the latter it is neutral. Experiments have shown that this phosphate exerts little influence upon bacterial development; in fact, it is less active than sodium chlorid. Hence, with litmus as an indicator, this compound reacts alkaline and prevents a proper addition of alkali, and as the amounts of phosphates vary in different lots of culture-media, the degree of reaction cannot be accurately controlled by means of litmus.

Mr. Fuller recommends "neutrality to phenolphthalein" as a "datum point" for reaction. The expression of the reaction in parts or in percentage is inaccurate and awkward. It is suggested that all reaction be expressed in terms of normal solution necessary to render one liter of the medium neutral to phenolphthalein. As, ordinarily, the media are acid, it is convenient to call the acid number plus, alkaline minus.

The procedure at Lawrence is to get all ingredients in solution, mix well, place 5 c. c. in a 15 cm. porcelain dish, add 45 c. c. of distilled water, boil three minutes, titrate with $\frac{n}{20}$ alkali, using phenolphthalein as an indicator, calculate the amount of normal alkali necessary to make the main bulk of the solution neutral, heat for the usual period, filter, and titrate again. The solution is now rather acid, and hydrochloric acid is added in such amount that to make one liter of the medium neutral to phenolphthalein 15 c. c. of normal alkali is necessary. This is for special work. In general work, using 1 c. c. of the water-sample to 5 c. c. of culture-medium, the acidity is increased one-fifth to offset the dilution.

The following table, condensed from Mr. Fuller's communication, shows very clearly how reaction influences the number of points of microbic life developed in any given sample, and confirms the observations on page 102.

C. C. OF NORMAL SOLUTION REQUIRED TO NEUTRALIZE ONE LITER OF THE CULTURE-MEDIUM.	POINTS OF MICROBIC LIFE IN ONE C. C.		
	Lawrence Sewage.	*Merrimack River Water.*	*Filtered Water.*
40	168,000	100	4
25	1,720,000	7,800	84
20	2,688,000	15,000	184
5	2,625,000	6,900	80
0	2,234,000	5,800	66
— 10	2,230,000	3,200	48
— 25	1,520,000	200	13

— indicates alkalinity to phenolphthalein.

As an illustration of the difficulty of recognizing a specific typhoid germ I may mention that, although there is every reason to believe that the Schuylkill water supplied to Philadelphia causes considerable typhoid fever, Dr. A. C. Abbott, of the Department of Hygiene of the University of Pennsylvania, informs me that he has never been able to find in that water the typical bacillus.

ANALYTIC DATA.

FACTORS FOR CALCULATION.

Parts per 100,000 $\times$ .7 = Grains per Imperial Gallon
" " 1,000,000 $\times$ .07 = " " " "
" " 100,000 $\times$ .583 = " " U. S. "
" " 1,000,000 $\times$ .058 = " " " "
" " 1,000,000 $\times$ .00833 = Pounds per 1000 U. S. Gal.
Grains " Imp. gallon $\div$ .7 = Parts per 100,000
" " " " $\div$ .07 = " " 1,000,000
" " U. S. " $\div$ .583 = " " 100,000
" " " " $\div$ .058 = " " 1,000,000

Parts per 100,000 divided by 2 and quotient increased one-tenth gives approximately grains per U. S. Gallon.

Al_2O_3, ... $\times$ .529 = Al
AgCl, ... $\times$ .247 = Cl
$BaSO_4$, ... $\times$ .588 = Ba
$BaSO_4$, ... $\times$ .412 = SO_4
B_2O_3, ... $\times$ .314 = B
CaO, ... $\times$ .714 = Ca
$CaCO_3$, ... $\times$ .40 = Ca
Cl, ... $\times$ 1.65 = NaCl
Fe_2O_3, ... $\times$ .7 = Fe
KCl, ... $\times$ .524 = K
$2KCl, PtCl_4$, ... $\times$ .16 = K
$2KCl, PtCl_4$, ... $\times$ .30 = KCl
$Mg_2P_2O_7$, ... $\times$ .218 = Mg
$Mg_2P_2O_7$, ... $\times$ .853 = PO_4
MnS, ... $\times$ .632 = Mn
NaCl, ... $\times$ .393 = Na
N, ... $\times$ 4.43 = NO_3
N, ... $\times$ 3.28 = NO_2
N, ... $\times$ 5.85 = $Ca(NO_3)_2$
NH_3, ... $\times$ .823 = N

CONVERSION TABLE.

Parts per Million.	Grains per U. S. Gallon.	Grains per Imp. Gal.	Parts per Million.	Grains per U. S. Gallon.	Grains per Imp. Gal.
1	.058	.07	26	1.508	1.82
2	.116	.14	27	1.566	1.89
3	.174	.21	28	1.624	1.96
4	.232	.28	29	1.682	2.03
5	.290	.35	30	1.740	2.10
6	.348	.42	31	1.798	2.17
7	.406	.49	32	1.856	2.24
8	.464	.56	33	1.914	2.31
9	.522	.63	34	1.972	2.38
10	.580	.70	35	2.030	2.45
11	.638	.77	36	2.088	2.52
12	.696	.84	37	2.146	2.59
13	.754	.91	38	2.204	2.66
14	.812	.98	39	2.262	2.73
15	.870	1.05	40	2.320	2.80
16	.928	1.12	41	2.378	2.87
17	.986	1.19	42	2.436	2.94
18	1.044	1.26	43	2.494	3.01
19	1.102	1.33	44	2.552	3.08
20	1.160	1.40	45	2.610	3.15
21	1.218	1.47	46	2.668	3.22
22	1.276	1.54	47	2.726	3.29
23	1.334	1.61	48	2.784	3.36
24	1.392	1.68	49	2.842	3.43
25	1.450	1.75	50	2.900	3.50

TABLE OF DISSOLVED OXYGEN.

Dibdin's Table of Oxygen Dissolved by Water at Various Temperatures, extended to give the Weight of Oxygen per Liter. Corrected to 0° C. and 760mm. Pressure.

Temperature Fahrenheit.	Temperature Centigrade.	Cubic Inches of Oxygen per Gallon (70000 Grains).	Milligrams of Oxygen per Liter.
41°	5.00°	2.101	10.84
42	5.55	2.074	10.72
43	6.11	2.048	10.57
44	6.66	2.022	10.45
45	7.22	1.997	10.30
46	7.77	1.973	10.18
47	8.33	1.949	10.06
48	8.89	1.927	9.94
49	9.44	1.905	9.83
50	10.00	1.884	9.72
51	10.55	1.864	9.61
52	11.11	1.844	9.51
53	11.66	1.826	9.42
54	12.22	1.808	9.33
55	12.77	1.791	9.24
56	13.33	1.775	9.15
57	13.89	1.760	9.08
58	14.44	1.746	9.01
59	15.00	1.732	8.94
60	15.55	1.719	8.87
61	16.11	1.706	8.80
62	16.66	1.695	8.74
63	17.22	1.683	8.68
64	17.77	1.674	8.64
65	18.33	1.667	8.60
66	18.89	1.660	8.56
67	19.44	1.652	8.52
68	20.00	1.644	8.48
69	20.55	1.639	8.45
70	21.11	1.634	8.43

The table is calculated for a barometric pressure of 760 mm., and would require corrections for variations from this, but such corrections are mostly within the limits of experimental error.

ANALYSES OF RAIN AND SUBSOIL WATERS.

Milligrams per Liter.

	From.	Total Solids.	Chlorin.	N by KMnO$_4$.	N as NH$_4$.	N as NO$_2$.	N as NO$_3$.
Rain water.	Bellefonte—collected by Prof. Wm. Frear, after long rain.	5	none	0.148	0.280	none	none
Subsoil water.	Wynnewood—pool fed by underground spring.	65	6.20	0.032	0.024	none	2.3
" "	Wynnewood—well about 150 yards from above.	60	4.00	0.024	0.016	none	3.5
" "	Wynnewood—well polluted by farm-yard drainage; about 500 yards from pool.	...	16.00	0.208	0.028	none	14.2
" "	Pump-well in densely populated district. Highly contaminated.	1120	57.00	1.00	3.120	0.01	33.0
" "	Newly dug well in populated district. Highly contaminated.	620	120.00	0.08	2.00	0.03	16.0
" "	Well at Barren Hill, 130 feet deep.	470	120.00	undet.	undet.	traces	22.0

RESULTS FROM SCHUYLKILL RIVER—MILLIGRAMS PER LITER.

Samples taken from hydrant at 715 Walnut Street.

1888	State of Weather.	Condition.	Total Solids.	Nitrogen by KMnO$_4$.	Nitrogen as NH$_4$.	Nitrogen as NO$_2$.	Nitrogen as NO$_3$.
Sept. 17.	Continued rain	Turbid	160	.06	.048	None.	.13
" 18.	Continued rain.	Muddy.	160	.09	.020	"	.34
" 19.	Clear.	Turbid.	140	.108	.004	Trace.	.34
" 20.	"	"	150	.124	None.	Faint trace.	.25
" 21.	"	Less turbid.	150	.132	"	None.	.25
" 22.	"	Very muddy.	180	.180	"	Faint trace.	.25
" 24.	"	Turbid.	140	.100	"	None.	.13
" 25.	"	Slightly turbid.	130	.092	.01	"	.13
" 26.	"	Slightly turbid.	120	.068	"	"	.18
" 27.	"	Very slightly turbid.	110	.056	None.	"	.18
" 28.	"	Very slightly turbid.	105	.052	"	"	.36
" 29.	"	Nearly clear.	125	.052	.008	"	.70
Oct. 1.	"	"	140	.052	.012	"	.70
" 2.	"	"	125	.044	None.	"	.40
" 3.	"	"	120	.080	0.6	"	.68
" 4.	"	Clear.	120	.084	.008	"	.80
" 5.	"	"	120	.092	.012	"	.70
" 6.	Rain.	"	120	.060	.012	"	.80
" 8.	"	Turbid.	150	.108	.008	"	.80
" 9.	Clear.	Slightly turbid.	130	.088	.008	"	.70
" 10.	"	Almost clear.	125	.056	.010	"	.70
" 11.	"	Clear.	115	.044	.010	"	.70

ANALYSES OF ARTESIAN WATERS.

Milligrams per Liter.

	PHILADELPHIA.						BALTIMORE.	INDIANA.
	9th and Chestnut Sts.	24th and Washington Ave.	15th and Walnut Sts.	Broad and Columbia Ave.	Broad and Chestnut Sts.	Black's Island.	Locust Point.*	Ft. Wayne.
Condition,	Clear.	Clear.	Slightly turbid.	Clear.	Clear.	Clear.	Clear.	Clear.
Reaction,	Alkaline.	Alkaline.	Alkaline.	Alkaline.	Alkaline.	Alkaline.	Alkaline.	Alkaline.
SiO_2,	36.00	34.50	35.00	24.00	29.00	19.50	4.00	...
SO_4,	61.80	36.26	9.90	25.84	15.24	102.36	4.10	...
PO_4,	0.24	0.31	0.31	Trace.	Trace.	0.38	...	...
CO_3 (combined),	64.91	110.17	54.80	78.84	108.64	50.64	...	...
Cl,	89.21	91.77	108.50	206.01	172.00	554.48	4.68	190.2
H_2S,	...	Trace.	...	...	...	...	...	...
Mn,	...	...	...	1.80	4.32	...	...	...
Fe,	9.80	7.70	5.70	5.28	...	1.05	...	...
Ca,	54.57	44.28	50.00	125.00	101.42	53.57	2.5	...
Mg,	15.27	7.98	12.60	30.70	22.27	6.48	0.86	...
Na,	54.13	84.13	32.46	72.34	55.83	378.63	...	...
K,	4.14	8.54	4.10	Traces.	6.28	Trace.	...	...
N by $KMnO_4$,	0.032	Not det.	0.048	0.048	0.015	0.148	0.012	0.035
" as NH_4,	0.248	Not det.	0.034	0.032	0.004	0.132	0.044	0.315
" " NO_2,	Trace.	None.	None.	0.02	Trace.	None.	None.	None.
" " NO_3,	Trace.	Trace.	None.	2.00	1.00	None.	None.	None.

* Mr. Wm. Glenn, Chemist of the Baltimore Chrome Works, kindly sent this sample.

EXAMPLES OF CITY SUPPLIES.

Milligrams per Liter.

City.	New York City.	Brooklyn.	Boston.	Cincinnati.	Chicago.	Little Rock, Ark.	Philadelphia.	Washington, D. C.
Class of Water,	Surface.	Subsoil.	Surface.	Surface.	Surface.	Surface.	Surface.	Surface.
Total Solids,	75.0	64.0	47.6 / 156.0	140.00	136.0	283.0	120.00	126.00
Chlorin,	2.8	13.5	5.1 / 34.8	14.00	2.3	8.2	4.00	2.8
Nitrogen as Nitrates,	0.28	16.0	0.07 / 0.5	0.26	Trace	1.1	1.00	1.0
Nitrogen as Nitrites,	None	None	0.001 / 0.01	...	None	...	None	.001
Nitrogen as Ammonium,	0.009	0.001	0.008 / 0.38	0.003	0.005	0.012	0.01	0.00
Nitrogen by permanganate,	0.06	0.07	0.16 / 0.23	0.09	0.075	0.13	0.10	0.12

The figures are taken from official reports and do not represent averages of large numbers of analyses, but individual analyses of samples representative of the supplies. The figures for Boston water represent the Cochituate and Mystic rivers respectively. The Little Rock supply is in the unfiltered condition; much of the solid matter is in suspension.

CULTURE-PHENOMENA OF SOME IMPORTANT MICROBES.

These data have been compiled principally from the writings of the Franklands (*Microörganisms in Water*), Vaughan (*Bact. of Drink. Water*, Proc. Amer. Ass'n Physicians, 1892), and Macé (*Tr. Prat. d. Bact.*). The references are indicated by appropriate initials.

In the first column + means "liquefies," O means "does not liquefy."
" second " + " "aërobic," — " "anaërobic."
" { third and fourth } " + " "grows," — " "grows feebly."
O " "does not grow."

The species marked * were described by Jordan (*Rep. S. B. of H., Mass.*, 1890), having been obtained from the sewage of Lawrence. For an account of recent study of microbes in Ohio-River water, see a paper by John W. Hill (*Trans. Amer. Soc. Civ. Eng.*, May, 1895).

Under the term *B. Aquatilis sulcatus*, Weichselbaum has described five varieties designated by the numbers I to V. All but V grow somewhat at blood heat.

CULTURE-PHENOMENA.

	Action on Gelatin.	Relation to Oxygen.	Action of Parietti's Solution.	Growth at Blood-heat.
Bacillus albus, V.,	0	+	0	0
" " anaërobus, V.,	0	—	0	0
" " putridus, V.,	+	+	0	0
" aquatilis sulcatus, F.,	0	+		+
" candicans,* F.,	0	+		+
" chlorinus, M ,	+	—		
" coli-communis, F. ,	0	+	+	+
" cinnabareus, V.,	+	+		0
" circulans, F.,	+	+		
" cloacæ,* F.,	+	+		+
" delicatulus,* F.,	+	+		+
" erythrosporus, F.,	0	+		
" figurans, V.,	+	+	0	—
" flavus, M.,	+	+		
" fluorescens liquefaciens, V.,	+	+	0	0
" " non-liquefaciens. V.,	0	+	0	0
" gasoformans, V.,	+	+	0	0
" gracilis aërobiens, V.,	0	+	0	0
" " anaërobiescens, V ,	0	—	+	—
" helvolus, V.,	+	+	0	—
" hyalinus, F.*	+	+		+
" invisibilis, V.,	0	—	+	+
" lactis aërogenes, F.,	0	+		0
" liquefaciens albus, V.,	+	+	0	0
" ochraceus, V.,	+	+	0	0
" prodigiosus, M.,	+	+	0	0
" reticularis,* F.,	+	+		+
" rubescens,* F.,	0	+		
" rubidus, V.,	+	+	0	0
" subflavus, V.,	0	+	0	0
" subtilis, M.,	+	+		
" superficialis, F.,	+	+		+
" tholeideum, F.,	0	+		+
" typhi abdominalis, F.,	0	+	+	+
" ubiquitus,* F.,	0	+	+	+
" venenosus, V.,	0	—	+	+
" " brevis, V.,	0	—	+	+
" " invisibilis, V.,	0	+	+	+
" " liquefaciens,	+	—	+	+
" violaceus,	0	+	0	0
Micrococcus aquatilis, F., V	0	+	0	—
" " albus, V.,	0	+	0	—
" " invisibilis, V.,	0	+	0	—
" " magnus,	0	+	0	—
" (diplococcus) aquatilis, V.,	0	+	0	0
" (streptococcus) "	0	+	0	0
" candicans, V.,	0	+	0	0
" " F.,	+	+		
" cereus, V.,	0	+	0	0
" luteus, V ,	0	+	0	0
" subflavus, V.,	0	+	0	0
Spirillum choleræ,	+	+		+
" Finkleri, M.,	+	+		+

INDEX.

ACIDS, action on lead, 108
—— Actinic method for organic matter, 50
Action of water on lead, 107
Aëration of water, 128
Agar-agar, 66
Albuminoid ammonia, 35, 96
Alkali carbonates, determination of, 82
Alkaline permanganate, 33
Allen, on boiler waters, 111
———, lead in water, 108
———, sulphuric acid in water, 111
———, test for zinc, 56
Alum, action of, 123
———, test for, 58
———, use of, 102
Aluminum, determination of, 58, 75
———, test for, 58
———, in scale, 112
Amido-naphthalene, 44
Ammonia, albuminoid, 35, 96
———, free, 34
———, free water, 31
———, from rain water, 95
———, process, 20
Ammonium chlorid, standard, 31
——— molybdate, 51
——— picrate solution, 42
Analysis, statement of, 89
Analytic operations, 21
Anderson and Ogsten, purification of water, 124
Antwerp water, purification of, 126
Artesian water, 13, 18
——— waters, composition of, 146
Arsenum, detection of, 56

BACHMAN'S method, 36
Bacilli species of, 149
Bacillus typhosus, culture of, 72
Bacteriologic examination, 61
Barium, detection and estimation of, 55
Barren Hill well, 20, 144
Barus, Carl, suspended matters in water, 15
Basin, platinum, 26

Bicarbonates in water, 129
Biologic examinations, 61
Black's Island well, 93, 146
Blarez' oxygen process, 51
Boiler mud, 111
——— water, 109
——— water, points to be determined in, 112
——— ———, purification of, 129
——— ———, statement of results from, 112
Boric acid estimation, 85
Bottle culture, 69
——— for test solution, 33
Burner, low temperature, 26

CALCIUM bicarbonate, 129
——— carbonate, action in boiler waters, 110
——— ———, solubility of, 112
——— compounds, removal of, 130
——— , determination of, 76
——— hydroxid for purifying water, 131
——— sulphate, insolubility of, 112
——— sulphate, action in boiler water, 112
Carbonates, action on lead, 109
———, determination of normal, 82
Carbon filters, 115
Carbonic acid, action of, 18
——— ———, free, determination of, 84
——— ———, effect on microbes, 71
Caustic soda, use of, in boiler water, 130
Cellar waters, examination of, 134
Chlorin, determination of, 28
———, significance of, 93
Chromium, detection of, 55
City supplies, 147
Clarifying water, 23, 114
Clark's process for purifying water, 130
Collection of samples, 20
Color comparator, 38
———, determination of, 23
———, significance of, 91
Comparison cylinders, 38

Control determination, 27, 77
Conversion of ratios, 142
Cooper, A. J., delicacy of tests, 61
Copper, detection of, 60
———— sulphate, standard, 60
———— zinc couple, 43
Corrosion of boilers, 110
Cultivation of microbes, 67
Culture-media, 67
———— phenomena, 149

DEEP water, 18, 93, 146
Demijohn for water samples, 21
Denitrification, 18
Dibdin, table of dissolved oxygen, 143
Distilled water, wholesomeness of, 92
Driffield, composition of boiler mud, 111
Drown, aëration, 128
Drown and Hazen, ignition of residue, 27
Drown and Martin, nitrogen determination, 35
Dupré, dissolved oxygen, 54, 100

FERROUS ammonium sulphate, standard, 52
Ferric sulphate, standard, 57
Filter paper, 23
Filters, Bischoff's, 116
————, Pasteur-Chamberlain, 116
————, sand, 118
————, spongy iron, 116
Filtration, 115
Fleck's silver method, 50
Fluorescein, use of, 132
Frankland, purification by precipitation, 122
———— and Tidy, standards of purity, 99
————, isolation of microbes, 73
————, nitrifying bacillus, 17
Fuller, G. W., culture-media, 137
———— sand filtration, 119

GALLON, Imperial, 89
————, U. S., 89
Gelatin culture-media, 67
————, liquefaction of, 70
Gérardin, dissolved oxygen, 100
Gill's method, 41
————, method for lithium, 80

HARDNESS, determination of, 81
————, permanent, 81
————, temporary, 81
Hard scale, 112
———— water, softening of, 130

Hehner's cylinders for color comparison, 38
————, method for hardness, 82
History of water, 13
Hunt, T. Sterry, water in rocks, 16
Hydrogen sulphid, titration of, 81

IDENTIFICATION of source of water, 132
Imperial gallon, 89
Interpretation of results, 89
Indol reaction, 71
Iodin, centinormal, 81
Iron, action of, in purification, 124
———— compounds, solution by water, 19
————, determination of, 57, 75
————, significance of, 93

KJELDAHL method, 38
Koch's culture method, 64

LACMOID, use of, 25
Lead, action of water on, 107
Lead, determination of, 59
————, nitrate, standard, 60
Leeds, actinic method, 50
————, dissolved oxygen, 100
Lime, purification of water by, 114, 130
Lithium compounds, use of, 132
————, detection of, 86
————, separation of, 80
Litmus, use of, 25
Locust Point well, 20, 146

MAGNESIA in boiler sludge, 111
Magnesium, determination of, 75
Magnesium chlorid, decompositions and effects of, 100
———— compounds, removal of, 130, 131
Magnetic carbid, 129
Mallet, ammonia process, 96
Manganese, detection of, 58
————, determination of, 76
Microbes, table of, 149
Mine water, 110

NAPHTHYLAMINE, 44
Nesslerizing, 34
Nessler reagent, 32
Nitric acid, diluted, 57
Nitrification, 17
Nitrates, action in boilers, 111
————, determination of, 41
————, formation of, 17
————, significance of, 98
Nitrites, determination of, 44
————, formation of, 17

INDEX.

Nitrites, significance of, 98
Nitrogen in ammonium compounds, 29, 94
——— as nitrites, 44
——— as nitrates, 41
——— by permanganate, 96
——, oxidation of, 16
———, total organic, 38

ODOR, determination of, 24
——— from residue, 28
———, significance of, 91
Organic matter, 16, 27, 90
——— ———, oxidation of, 50
———, precipitation of, 123
Oxygen consumed, 46
——— -consuming power, 46
———, amount of, dissolved, 143
———, dissolved, determination of, 51
———, dissolved, effects of, 100, 110

PARAMIDOBENZENE - Sulphonic acid, 44
Parietti's solution, 73
Pasteur-Chamberlain filter, 116
Permanganate method, 46
——— standard, 47, 49
Pettenkofer's method for free carbonic acid, 84
Phenol broth, 73
——— disulphonic acid, 41
Phenolphthaleïn, use of, 25
Phosphates, action on lead, 109
———, determination of, 51
———, significance of, 94
———, use of, in purifying water, 131
Plate culture, 67
Platinum, preservation of, 25
Porter's process for purifying water, 131
Potassium, determination of, 79
——— chromate solution, 28
——— iodid solution, 41
——— nitrate, standard, 47
Potassium permanganate, alkaline, 33
——— permanganate, decinormal, 52
Potato culture, 66
Preliminary examination, 21
Pure water, corrosive action of, 109
Purification of boiler water, 129
——— of drinking waters, 114

RAFTER'S method, 62
Rain water, 13, 95, 144
Ratios, conversion of, 141, 142
Reaction, 24
Residue, charring of, 27
Results, statement of, 89

River water, 13, 14, 91, 145, 147
Roll culture, 69

SALT, action on boiler waters, 111
Samples, collection of, 21
Sand filters, 118
Sanitary application, 90
——— examinations, 21
Scale, 111
Schuylkill River water, composition of, 14, 145
Sedgwick's method, 62
Sewage, action of, 18, 98
Silica, action of, in water, 107
———, determination of, 75
——— in scale, 112
Silicates, action on lead, 107
Silver nitrate, standard, 28
——— ———, test in Porter's process, 132
——— nitrite, preparation of, 45
——— test for organic matter, 50
Sludge, 111
Solids total, determination of, 25
Solids, significance of, 92
Sodium and potassium, separation of, 79
——— carbonate, solution of, 31, 39
——— ———, standard, 81
——— ———, use of, in boiler waters, 130
——— chlorid, standard, 28
———, determination of, 79
——— hydroxid, solution, 39
——— nitrite, standard, 45
——— thiosulphate, 47
Source of water, tracing of, 132
Smart, C., nature of organic matter, 96
Specific gravity, 87
Spectroscope, 85
Spectroscopic examination, 85
Spongy iron filters, 116
Starch indicator, 48
Sterilizer, 65
Storage, effect of, 122
Subsidence, promotion of, 15, 122
Subsoil water, 15, 144
Sulphanilic acid, 44
Sulphids, formation of, 19, 92
Sulphuretted hydrogen, determination of, 81
Sulphuric acid, diluted, 49
——— ———, standard, 81
Surface water, composition of, 13, 14, 145, 147
Suspended matters, 14, 15

TASTE, significance of, 91
Technic examinations, 109
Tests for metals, delicacy of, 61

Tidy's permanganate process, 47
Tidy and Frankland, standard of purity, 99
———, Odling and Crookes, lead in water, 107
Tri-sodium phosphate, use of, for purification, 131

UNCONTAMINATED waters, 101
Unit standard, 136
Urea, decomposition of, 95
Urine in water, 95
U. S. gallon, 89

VAUGHAN, typhoid fever germ, 107
Vegetable matter, 28, 96
Vegetable growth in water, 62

WATER, amount of, in rocks, 16
Wanklyn, standards of purity, 97
Whipple, unit measure, 136
———, growth of microscopic organisms, 137
Wurtz' lactose-agar, 137

ZINC, detection of, 56

www.ingramcontent.com/pod-product-compliance
Lightning Source LLC
Chambersburg PA
CBHW030320170426
43202CB00009B/1085